An Atlas of Human Dental Vascularity and Innervation

An Atlas of
Dental Vascularity
& Innervation

Craig Barrington, DDS

Private Practice
Waxahachie, Texas

QUINTESSENCE PUBLISHING

Berlin | Chicago | Tokyo
Barcelona | London | Milan | Mexiko City | Moscow | Paris | Prague | Seoul | Warsaw
Beijing | Istanbul | Sao Paulo | Zagreb

Library of Congress Cataloging-in-Publication Data

Names: Barrington, Craig, author.
Title: An atlas of dental vascularity & innervation / Craig Barrington.
Other titles: Atlas of dental vascularity and innervation
Description: Batavia, IL : Quintessence Publishing, [2021] | Includes bibliographical references. |
  Summary: "This book documents the internal anatomy of teeth, with a special focus on the innervation
  and vascular structure and distribution within the dentin chamber. Instead of the common use of
  sectioning, the author uses a complex diaphanization method to make extracted teeth transparent
  before photographing the intact internal dental anatomy. Therefore, the images throughout the book
  display structures that have rarely been seen so clearly and in three dimensions, including clear
  photographs of the pulp chamber, apical anatomy, and tooth channels. The body of work represented in
  this book pushes our understanding of internal dental anatomy"— Provided by publisher.
Identifiers: LCCN 2021030983 | ISBN 9781647241001 (hardcover)
Subjects: MESH: Tooth--anatomy & histology | Atlas
Classification: LCC QP88.6 | NLM WU 17 | DDC 612.3/11--dc23
LC record available at https://lccn.loc.gov/2021030983

A CIP record for this book is available from the British Library.
ISBN: 9781647241001

QUINTESSENCE PUBLISHING
USA

© 2022 Quintessence Publishing Co, Inc

Quintessence Publishing Co, Inc
411 N Raddant Road
Batavia, IL 60510
www.quintpub.com

5 4 3 2 1

Editor: Bryn Grisham
Design & production: Sue Zubek
Printed in Croatia

*For my parents who have never stopped supporting
me and who have always had faith in me.*

*I would also like to thank my brothers, my wife, and my kids
for their continued support and feedback.*

*There is a God. There has to be. There is no
other explanation, as far as I can tell.*

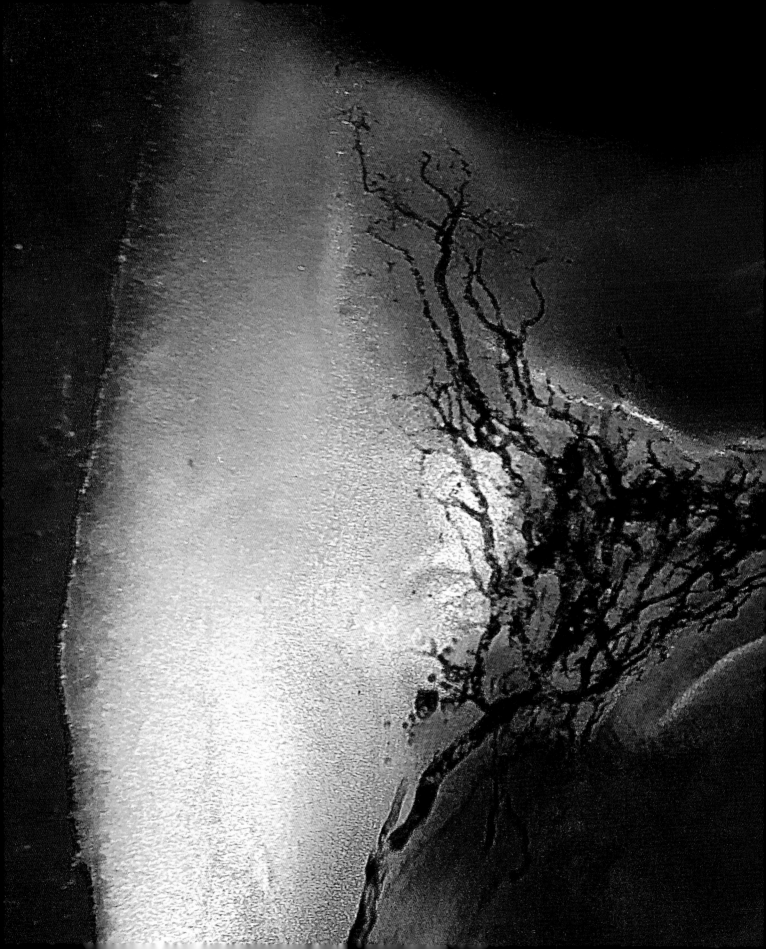

Contents

Preface

"Respect for life requires that science and technology should always be at the service of man and his integral development. Society as a whole must respect, defend, and promote the dignity of every human person, at every moment and in every condition of that person's life."

—*Pope John Paul II, Evangelium Vitae*

Over the years, when I have shown some of these images to people, the reactions are universally profound. Most of the time, I get a comment to this effect: "These are works of art, you should sell them!" It has been my thought that to sell or profit from these images would be inappropriate or even unethical. Even if these photographs are considered works of art, they are still images of human extracted teeth. It might be an exaggeration but I see it as akin to some form of human trafficking of human body parts.

Secondarily, the person that these extracted teeth came from should be given the opportunity share in any profits I would receive from selling pictures. However, there is only one case in this book in which I know the patient to whom the tooth actually belongs. Otherwise, I have no clue to whom the extracted teeth belong, and there would be virtually no way to trace, track, or attach the results to the human donor.

I have looked into the ethics regarding what I have accomplished, in terms of revealing the internal anatomy of extracted human teeth. I have made phone calls to the sources listed by the American Dental Association, spoken with leaders at dental schools, read books, and consulted with many people. Repeatedly, the answers I get are "keep going" or "you're fine, they're just teeth." With that, I have continued to work toward better and better results, knowing one day that this topic would need to be approached.

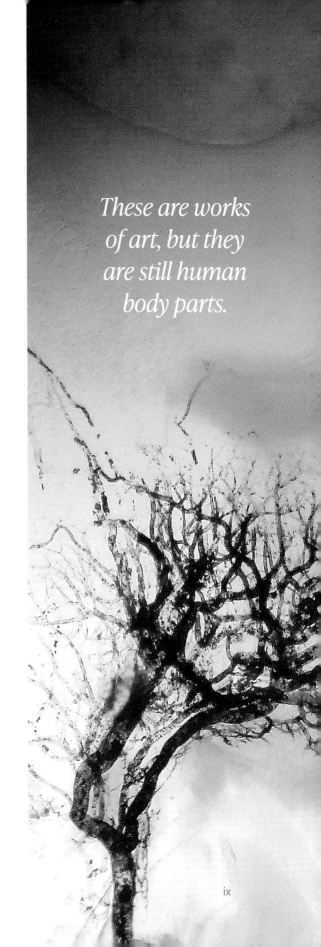

Thus far, all of my research has taken place in the confines of my privately owned dental office. I have paid for all of the supplies with my own personal money. I have accomplished all of the work, results, and innovation in photography techniques completely on my own, unassisted by any outside source. In order to gain a quick volume of samples, specimens were gratefully acquired from oral surgeons and oral surgery departments. I have in no way, fashion, form, or manner profited from the results shown here.

We are missing a lot in dentistry. We extract and discard human teeth like they are trash. No other field of medicine trashes and discards human body parts at the same rate and with the same disregard as found in the field of dentistry. It would be my hope that the anatomical images featured in these pages would change this and set us on a new path, not only to save teeth but save them with a more educated approach than what is prescribed by our current profession.

These are works of art, but they are still human body parts.

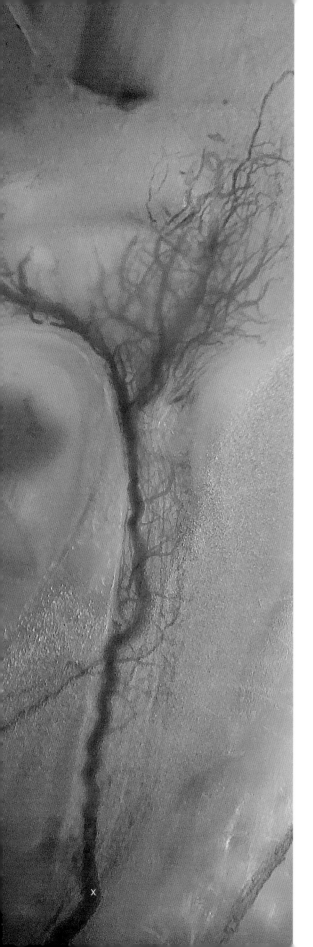

Acknowledgments

I would like to thank Dr Stephanie Roberts Jackson for taking the time to go through this publication systematically and provide amazing feedback. She helped to make this a better and more educationally valuable publication. She was helpful and patient beyond anything I could expect. Dr Roberts Jackson, I always appreciated your realistic and sound perspective from my years in dental school, and I am so pleased to know that you are still that amazing person more than 20 years later. Thank you!

Introduction

This compilation represents almost two decades of a journey in curiosity and amazement. It has been an absolute privilege to be guided to this path. Progress through continued failure is the only explanation for the results that have been compiled in this book. "Fail fast; fail forward" was my key quote and mantra. Failure became so beneficial that I found myself almost unable to wait for the next experiment to mess up. It was with deep, calculated observation of patterns and the next available aberration that brought me forward to the next breakthrough. I have literally not worked a day of my life on this project. It has all been energy based in the pure joy and amazement that I have found in the discovery of human anatomy and the associated revealing processes and history of diaphanization.

There are no methods revealed in the context of this publication. The methods have not been written down or stored.

There is not a lot of information available on the vascularity and nerve distribution within a human tooth. Most—if not all—pictures of the vascularity and innervation of the human pulp are artist-generated depictions of what they "believe" or even "wish" is inside our teeth. Other documentations are rare and potentially not easily repeated. Most others are on animal specimens other than humans. There is certainly more than we think hiding inside each and every single tooth. There are literal miles of blood vessels and structures that are not well documented. Given the lack of actual documentation, the structures, functions, and limitations of these tissues are not even close to being understood. Histologic sections slice and dice through any potential 3D rendering, which limits our understanding. It has been estimated that if all the blood vessels in an adult were laid out in a single line, they would go around the earth four times. My guess is that this publication increases that number significantly.

Diaphanization has been around for a long time. Werner Spalteholz is the denoted founder and father of all methods of making transparent whole tissue sections or entire organs. I studied in depth his method of diaphanization for whole organs and then scientifically broke down each step over 18 years until I understood every single nuance of every single step. With that, I was able to develop faster and more definitive methods than previously used. Human teeth pose a very unique and frustrating problem in that they are a mineral (enamel), then a combined mineral and tissue (dentin), and then a tissue (pulp). The dental pulp is an amazing and complex structure, as you will appreciate fully in the pages of this text, and it has been an honor to witness its grandeur, live and in person.

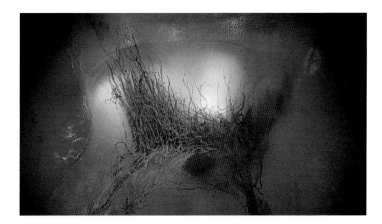

THE COVER IMAGE

This image is an extracted mandibular first molar from a person who is about 55 years old. This photograph demonstrates the internal vascularity and vascular distribution within the pulp chamber of that tooth. Specifically, the focus is on the mesial buccal pulp horn. This was photograph #503 of a series acquired over a 2-week period of working with this sample. I would not consider myself a photographer. Through the microscope, I could see the details you can see in this photograph, but I struggled to get the camera to attain the details shown in the final picture. Perimeter lighting via the aid of three prisms and continued experimentation with various light sources finally allowed the camera I used to capture what is shown. To say the least, it is a complete honor to be able to present this to the world. Minor enhancing and sharpening of the image was performed in a photo-editing program; otherwise, this is a virtually raw image. I have been challenged in the past regarding why I do not use focus stacking of several images to increase the depth of field. The answer is that it has simply been my personal preference to maintain images that are as close to a raw as possible, while highlighting specific aspects.

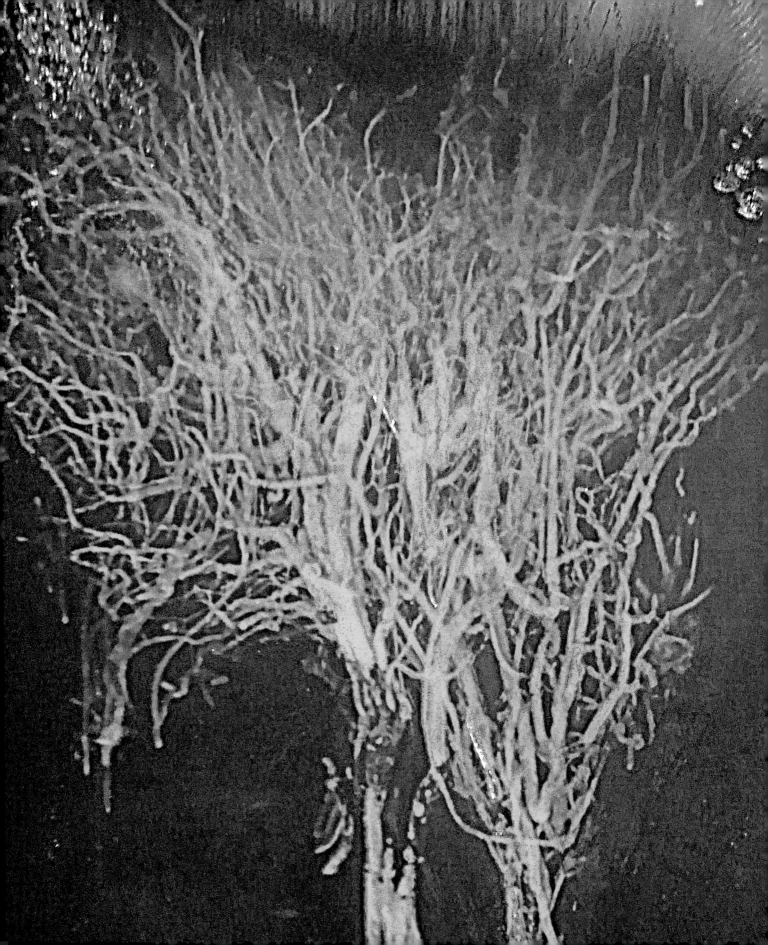

1

The First Results

Opportunity and use of the internet are key elements in the discoveries demonstrated in this book. Several years ago, I attended a dental continuing education course on a Saturday in which some intriguing information was discussed. The speaker offered references that did not seem to agree with what was said in the presentation. This caught me by surprise, and I started to check the listed articles more carefully. Oddly enough, the information I found in the references definitively contradicted what was presented in the lecture. This mismatch was my impetus. It sent me searching even further, only to discover products as well as techniques that gave me some amazing ideas that I wanted to put into action in my own lab immediately. I also found by accident key components (almost "household" items) that I was also able to purchase right away and have delivered to my office by Monday morning. All of it came together quickly. Years ago, all of these activities would have taken months or even years, whereas I was able to complete my initial plan and move it forward within hours.

A patient was scheduled for an extraction that very Monday morning. I quickly assembled all of the purchased items, and by implementing a few ideas and making some good guesses, I was able to acquire one of the most pristine samples to date, as shown on the recto. A lot of mystery, luck, conjecture, and even an arguable miracle were necessary for me to acquire the pictured result. When the vascularity inside the tooth revealed itself, I sat in total, absolute, and overwhelmed disbelief. A door had been opened, and I saw the possibilities.

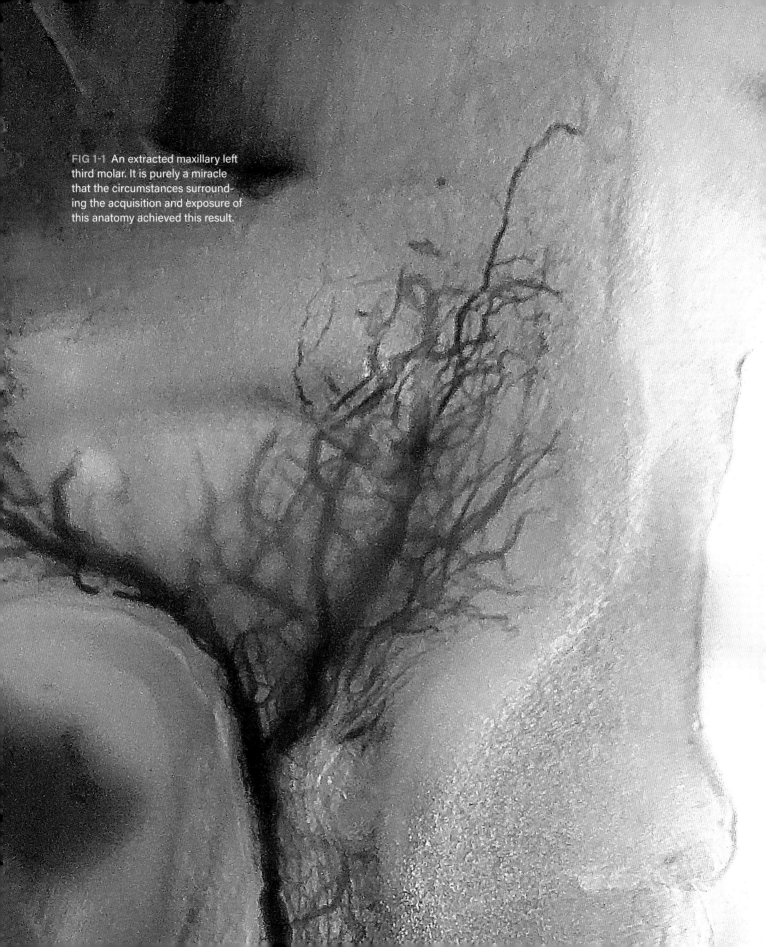

FIG 1-1 An extracted maxillary left third molar. It is purely a miracle that the circumstances surrounding the acquisition and exposure of this anatomy achieved this result.

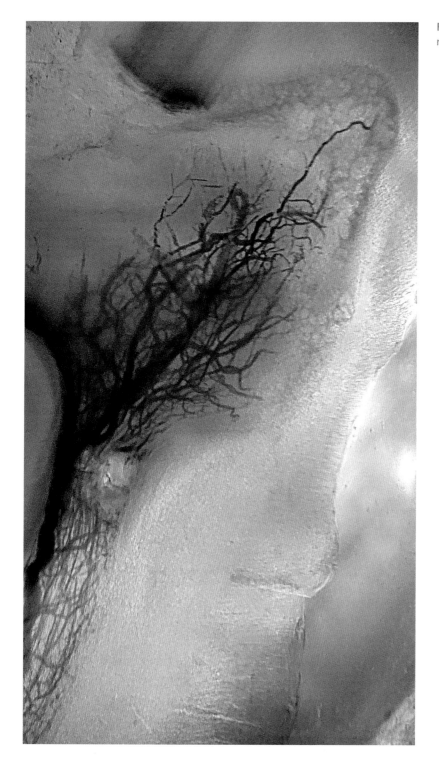

FIG 1-2 The hole seen to the right bottom is an error.

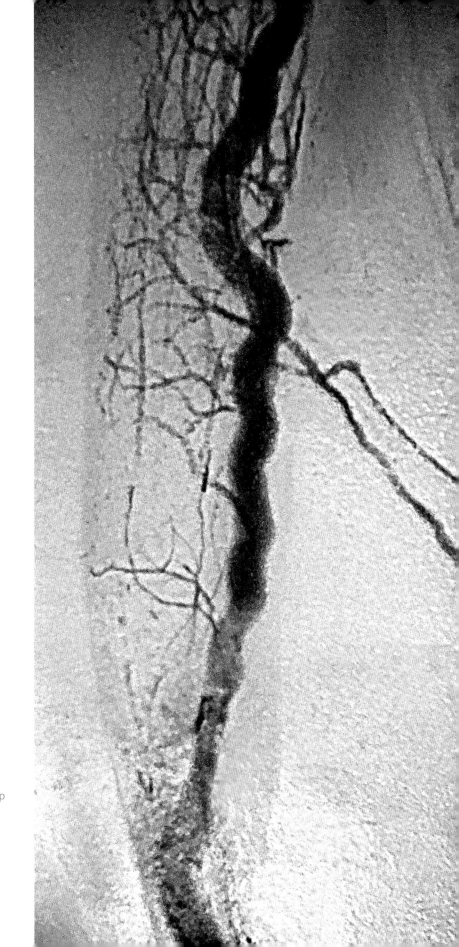

FIG 1-3 A closer view of the vascularity of the pulp channel system.

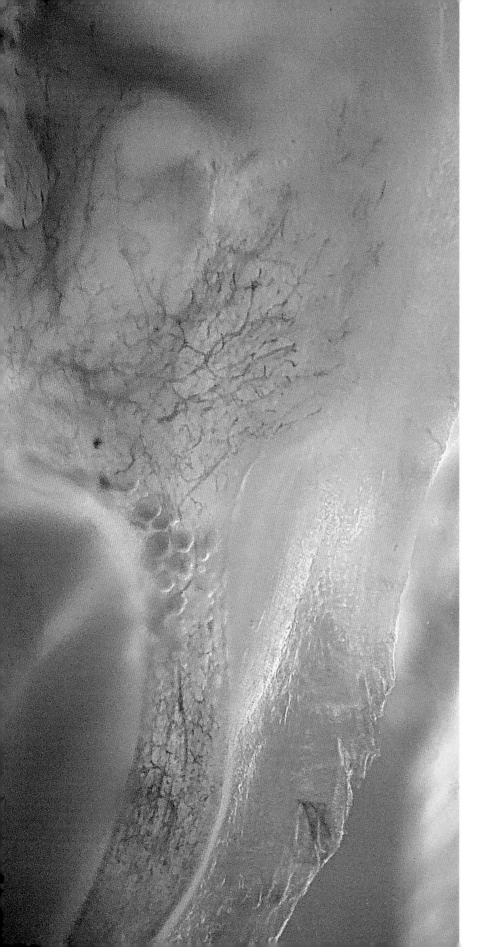

An impressive series of denticles are witnessed at the channel orifice. This is the first sample to demonstrate pulp aberrations that have been spoken of and witnessed during the clinical endodontic procedure but never routinely demonstrated or witnessed in histology. Odontoblastic processes decrease in number from chamber to apex, so it is interesting that these denticles are concentrated at the channel orifice.

FIG 1-4 The second result achieved. A large pulp stone is seen in the center of the chamber.

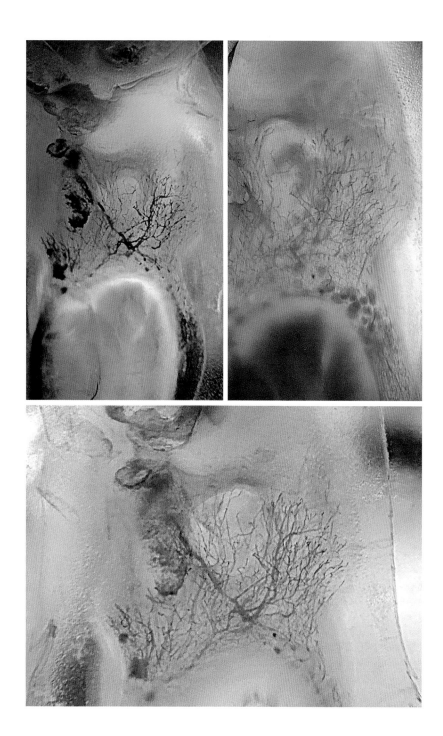

Further Reading

Lo Giudice G, Cutroneo G, Centofanti A, et al. Dentin morphology of root canal surface: A quantitative evaluation based on a scanning electronic microscopy study. Biomed Res Int 2015;2015:164065.

2

Varied Outcomes and Exposures of the Vascularity of Human Teeth

This chapter is a conglomerate of diaphanized imaged outcomes. The results vary from some early not-so-impressive but viable outcomes to what represent some of the most thorough and detailed examples ever seen. The underlying message of this collection of images is that revealing the contents of a human dental pulp chamber is a complex problem. To visualize the contents of the pulp chamber requires exquisitely detailed and well-timed steps and an acute attention to detail. It also requires an automatic acceptance of frustration and failed outcomes. Overall, it is an artful setup that has not (up to this point) been accomplished repeatedly nor well documented in dentistry. The more commonly seen outcomes result in sectioning of the tooth, sectioning of the pulp tissue, or simply removing and sectioning of the pulp all together. Most of these approaches lose the relative value of the dentin chamber walls; any sort of 3D view is lost by these methods.

Over years of trial and error, I have overcome the listed obstacles. I have found processes that not only expose the internal structures of the tooth chamber but also maintain the dentin structure. This chapter shows these results. The main focus is the vascular structure and distribution within the dentin chamber of a human tooth. This anatomy has been the hardest to demonstrate, as it is the most protected by varied contrasting layers of tissue. These structures have never been revealed in this much detail or with this much repetition and comparison. Patterns are important, and although this chapter shows significant variation—and in some cases error— enough is present to demonstrate a reasonable sampling of human dental vascular distribution with the anatomical chamber of a tooth.

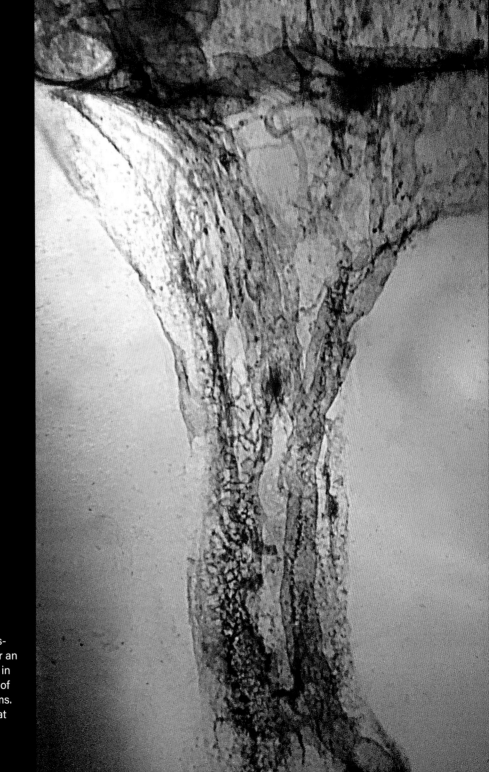

FIG 2-1 This is an early attempt to expose the vascularity of teeth stored for an unknown amount of time in unknown concentrations of unknown storage mediums. It was quite surprising that any sort of anatomy was able to be exposed.

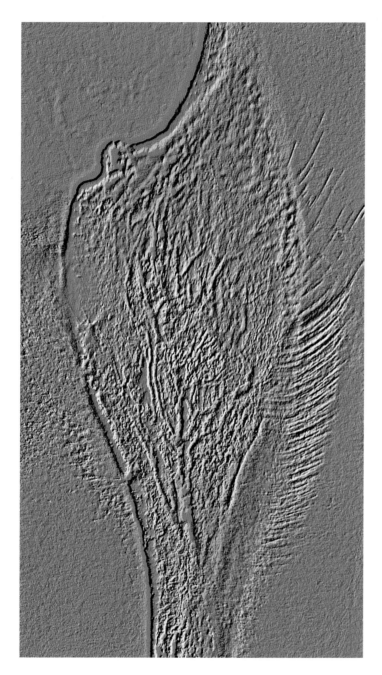

FIG 2-2 An attempt at exposure of pulp vascular demonstration via topographic methodology.

FIG 2-3 (right) A topographic demonstration of the pulp chamber contents. Vascularity and nerve plexus are present.

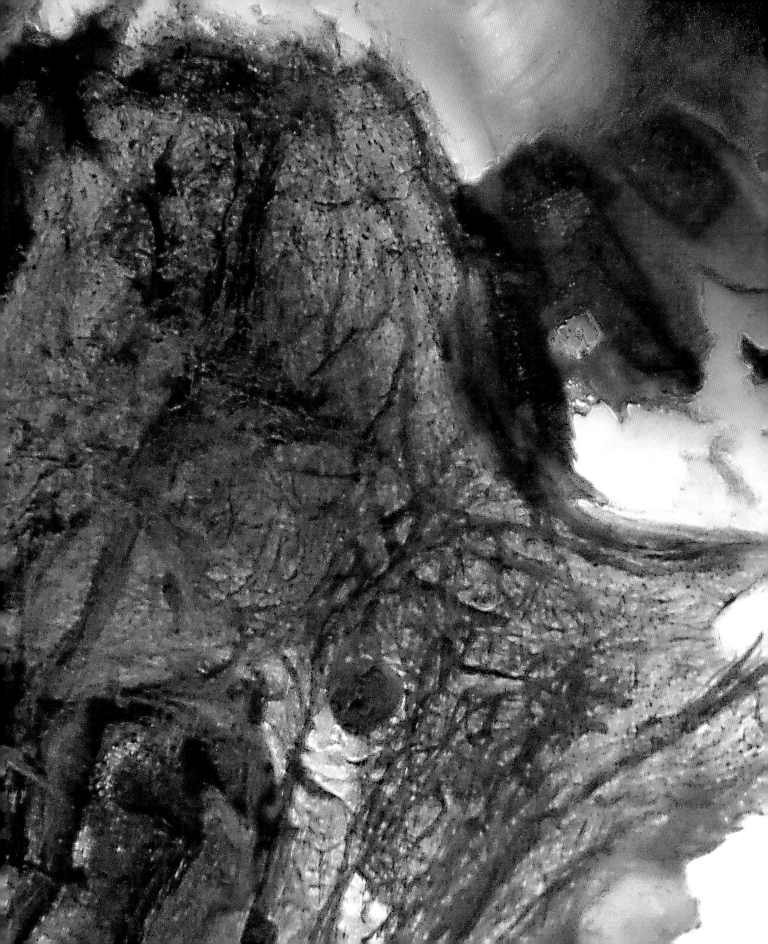

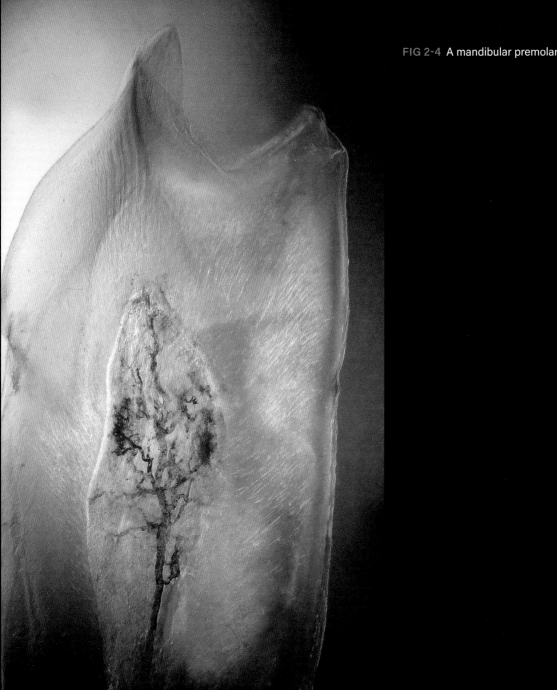

FIG 2-4 A mandibular premolar

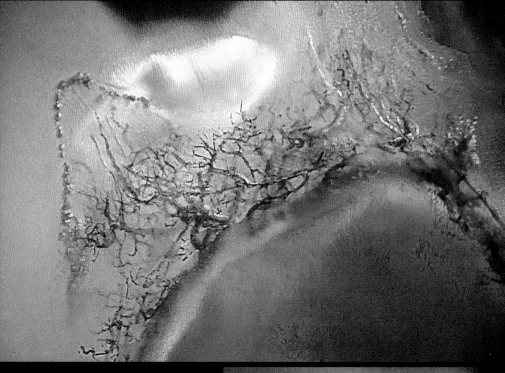

FIG 2-5 Chamber and closer look at the pulp horn of a maxillary molar.

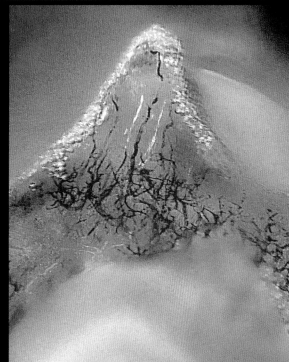

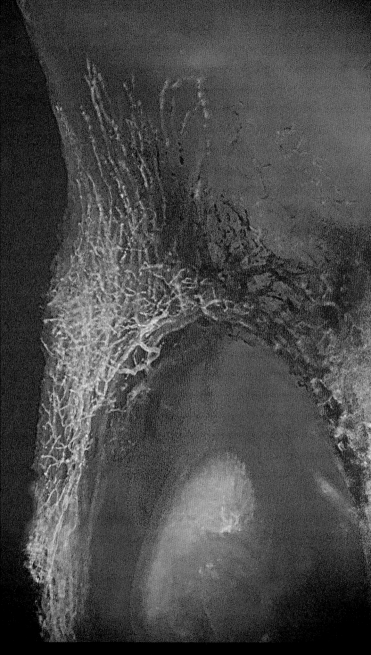
FIG 2-6 Vascularity of a mandibular molar.

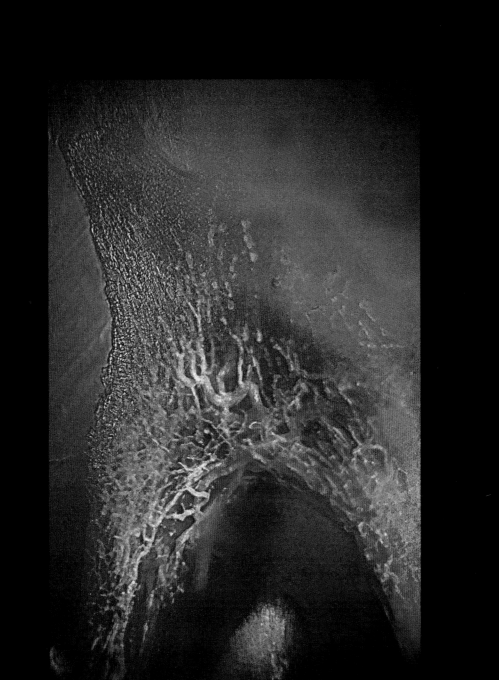

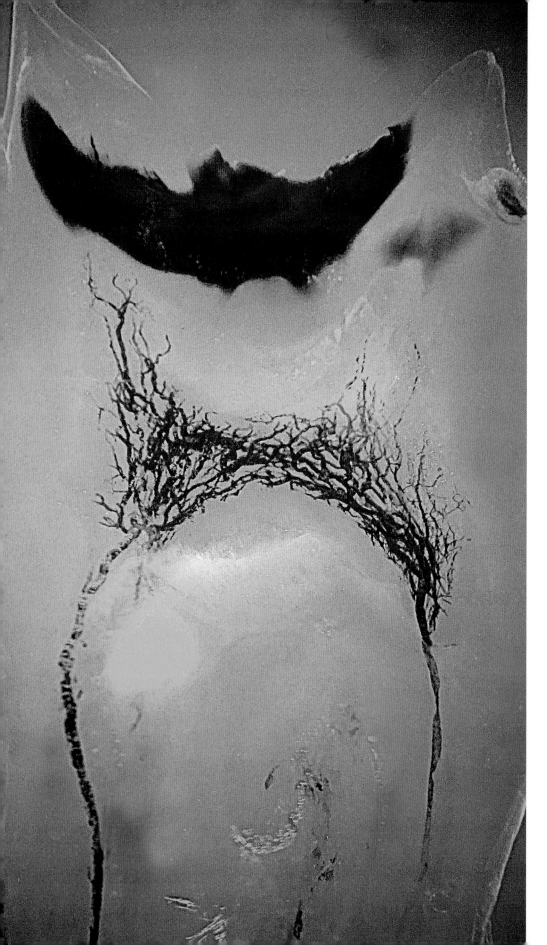

FIG 2-7 A maxillary molar. One of the most complete and thorough exposures to date. The *purple* above the chamber and channel system is decay.

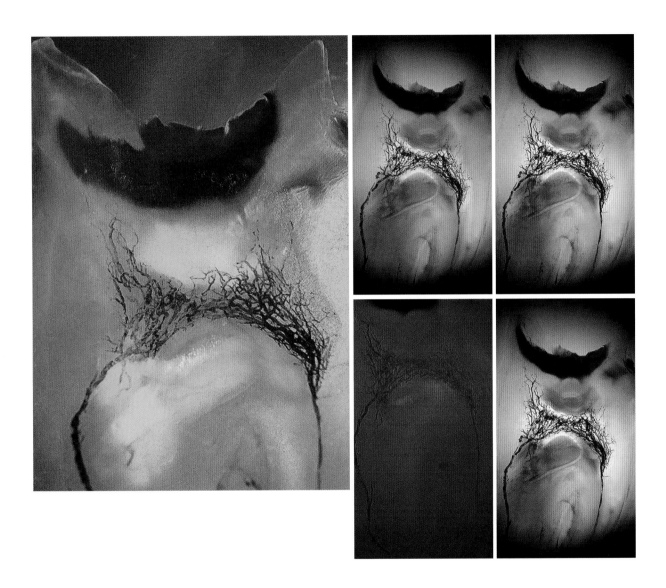

FIG 2-8 A maxillary molar.

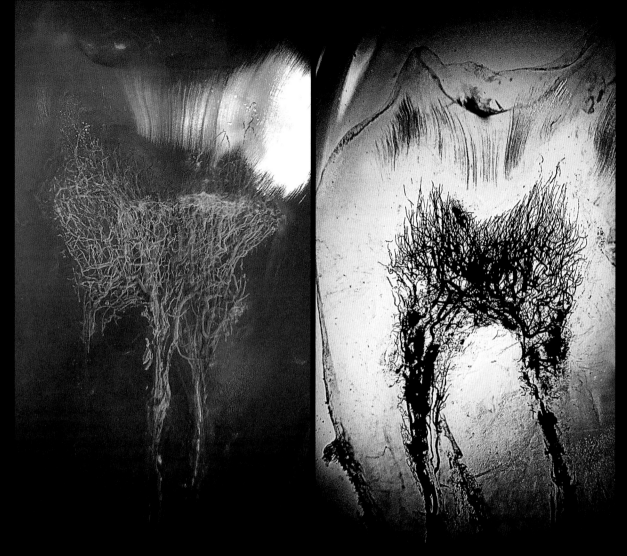

FIG 2-9 Two different maxillary molars.

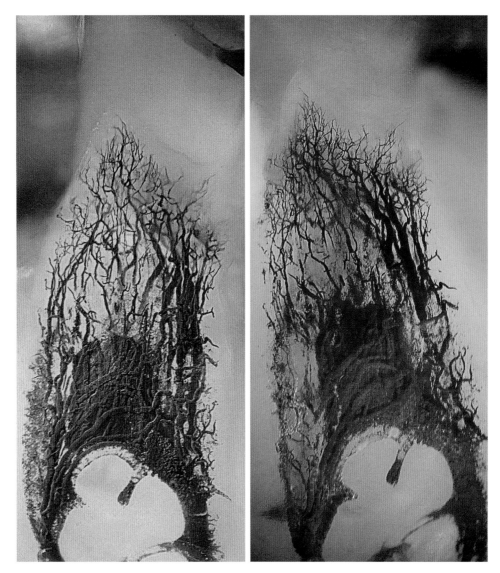

FIG 2-10 A mandibular second molar.

FIG 2-11 Massive vascularity along the channel and chamber wall of a mandibular second molar.

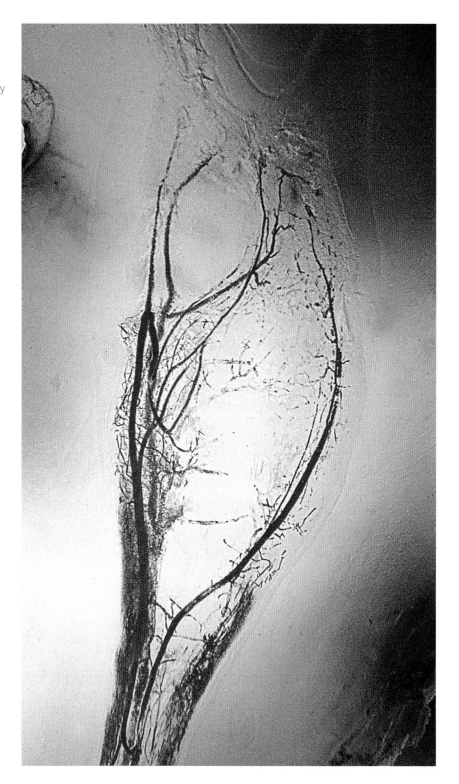

FIG 2-12 A maxillary second premolar.

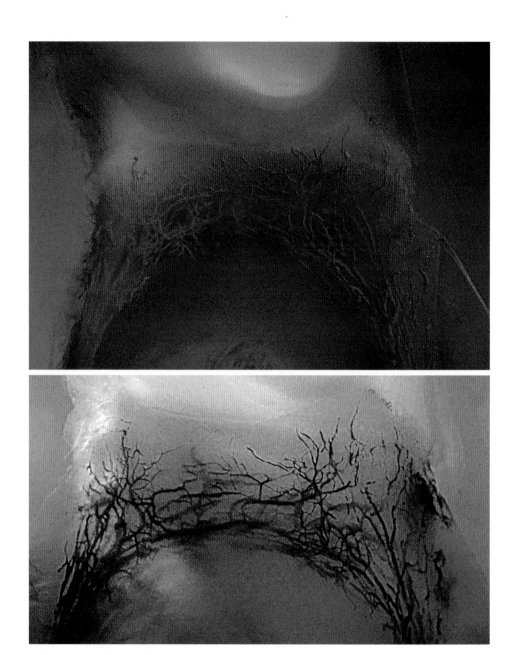

FIG 2-13 Varied perspectives showing the chamber anatomy of a mandibular molar.

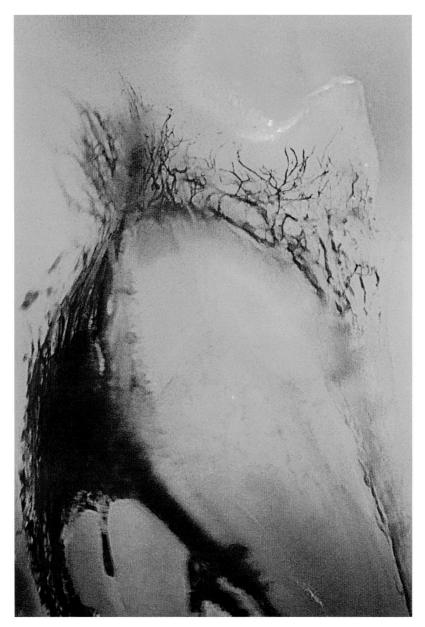

FIG 2-14 A mandibular second molar.

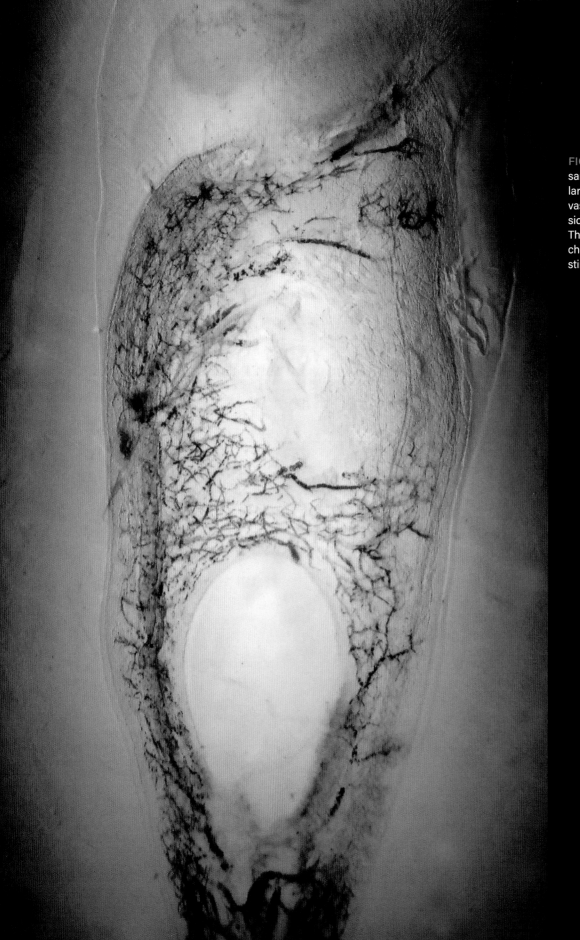

FIG 2-15 Images of the same maxillary premolar demonstrating denser vascularization to the palatal side versus the buccal side. This potentially relates to chemical/physical repeated stimulation from the buccal.

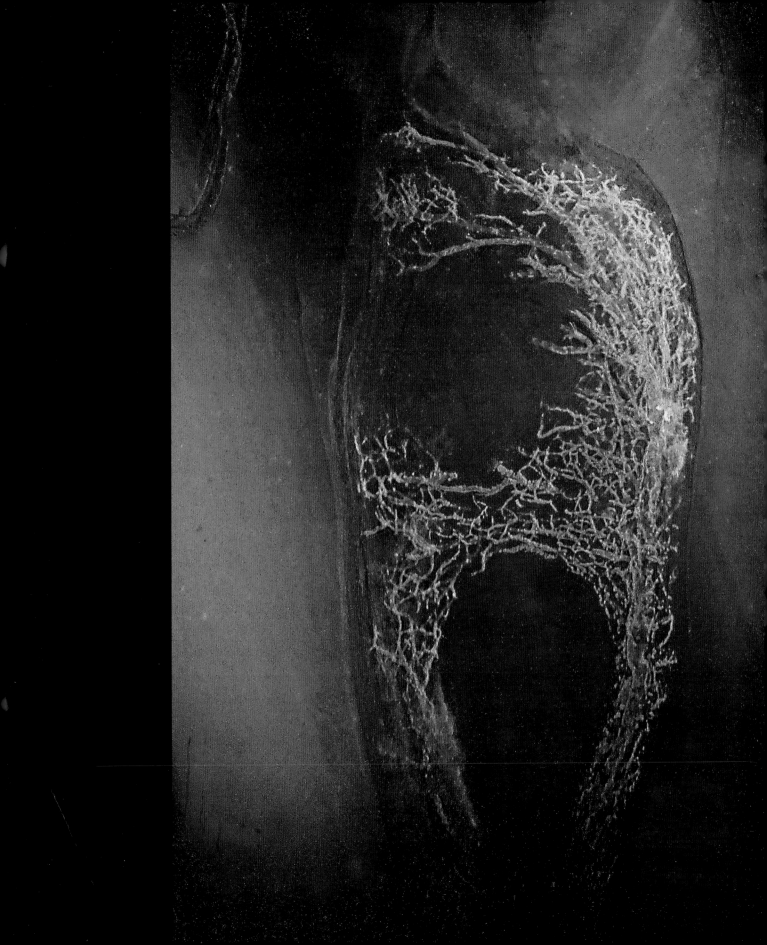

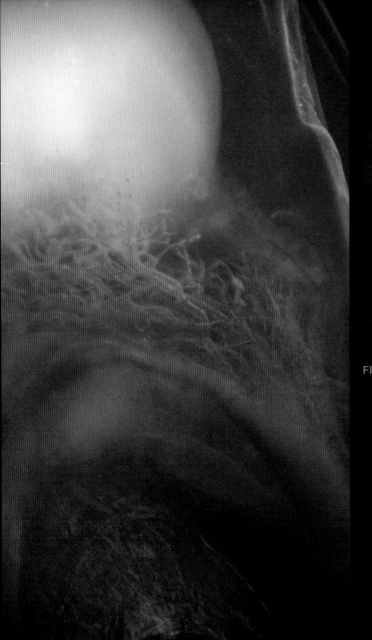
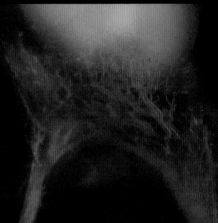

FIG 2-16 A mandibular molar.

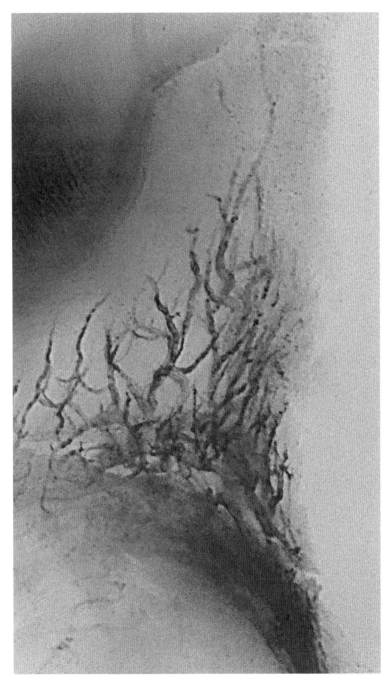

FIG 2-17 A closer look at the pulp
horn of the mandibular molar.

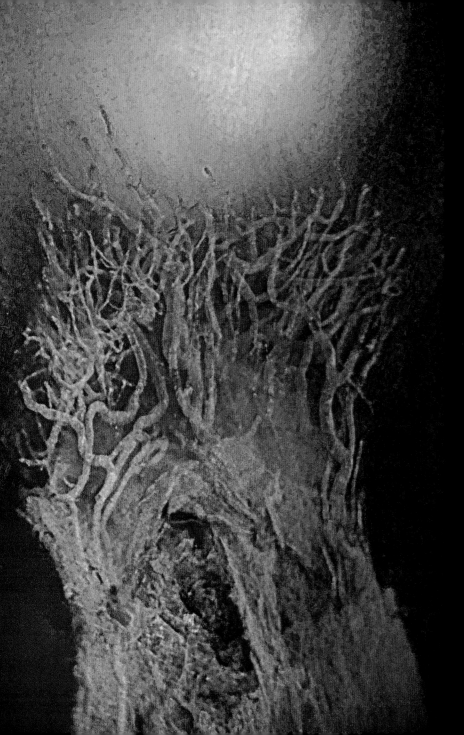

FIG 2-18 A mandibular molar, mesial and side views.

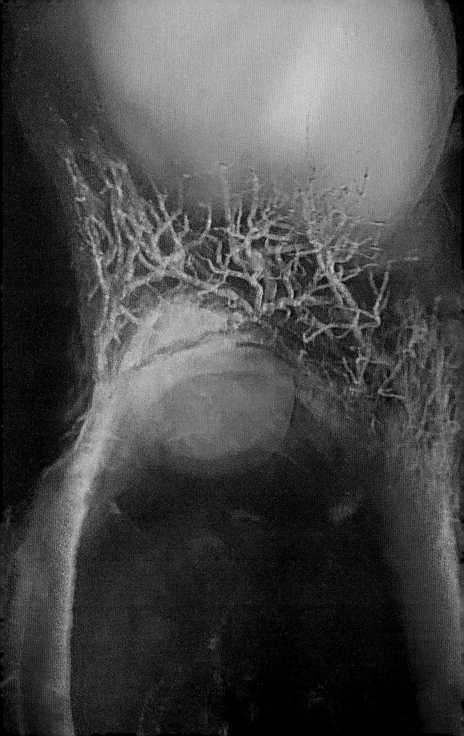

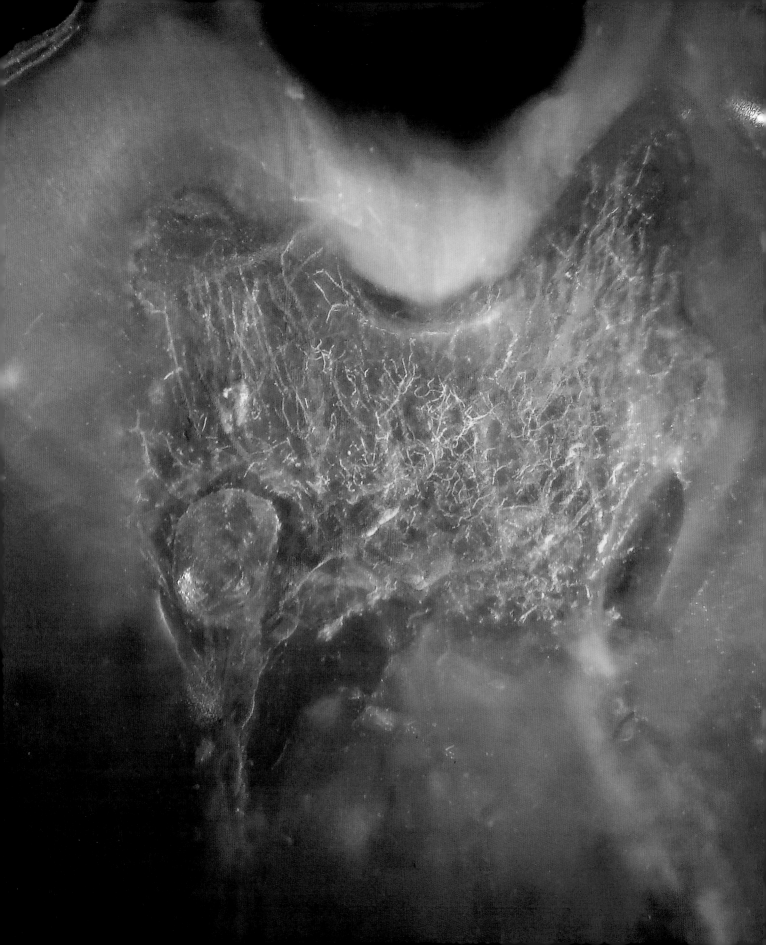

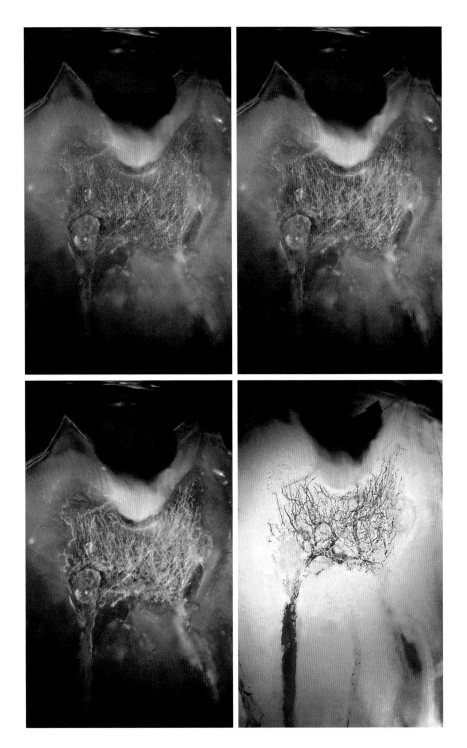

FIG 2-19 A pulp stone at the palatal channel orifice of a maxillary molar. Various views of the pulp stone and vascularity of pulp chamber of the maxillary molar.

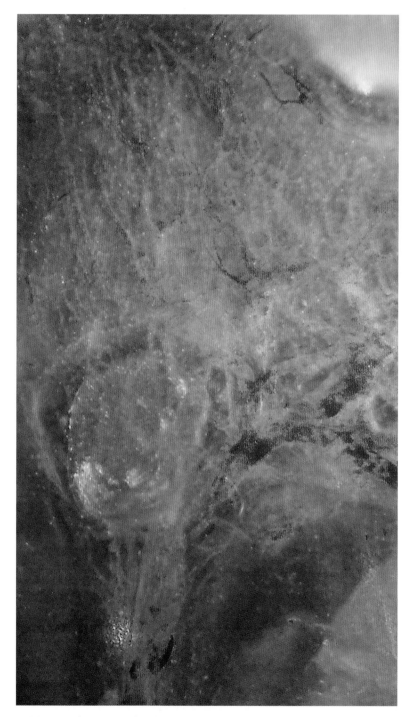

FIG 2-20 Pulp stone at the channel orifice.

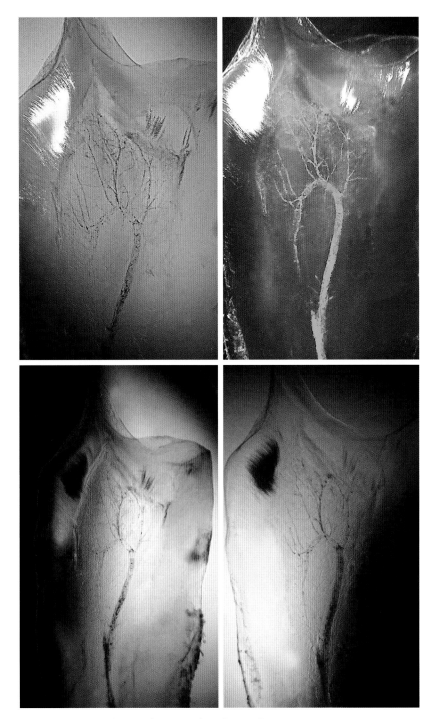

FIG 2-21 Vascular looping is a recurring observation in teeth with varied degrees of pulpitis.

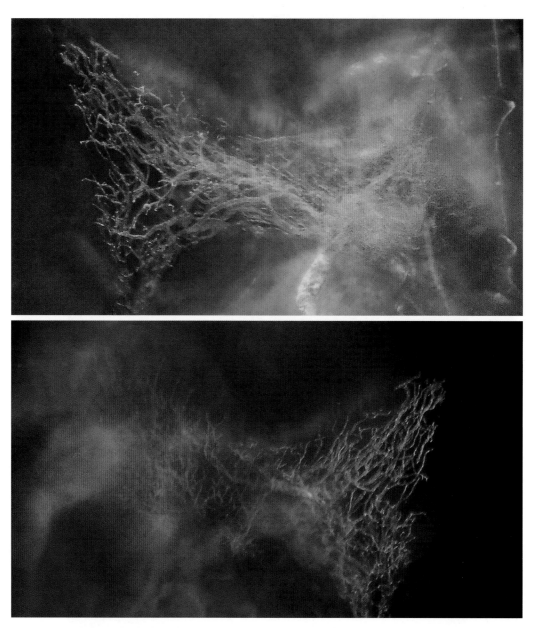

FIG 2-22 A maxillary molar.

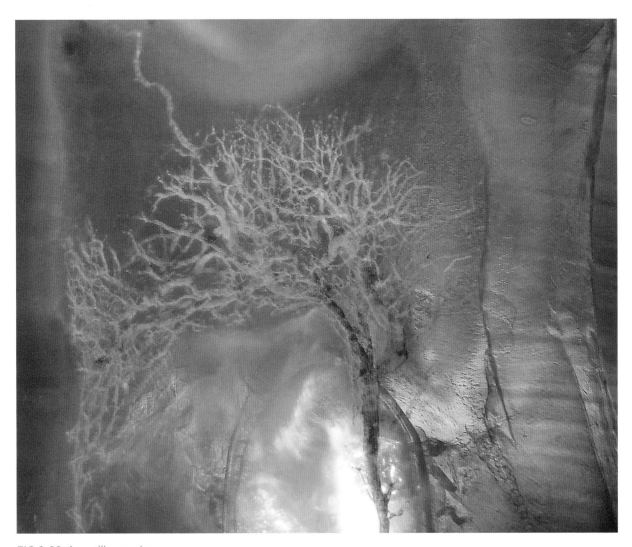

FIG 2-23 A maxillary molar.

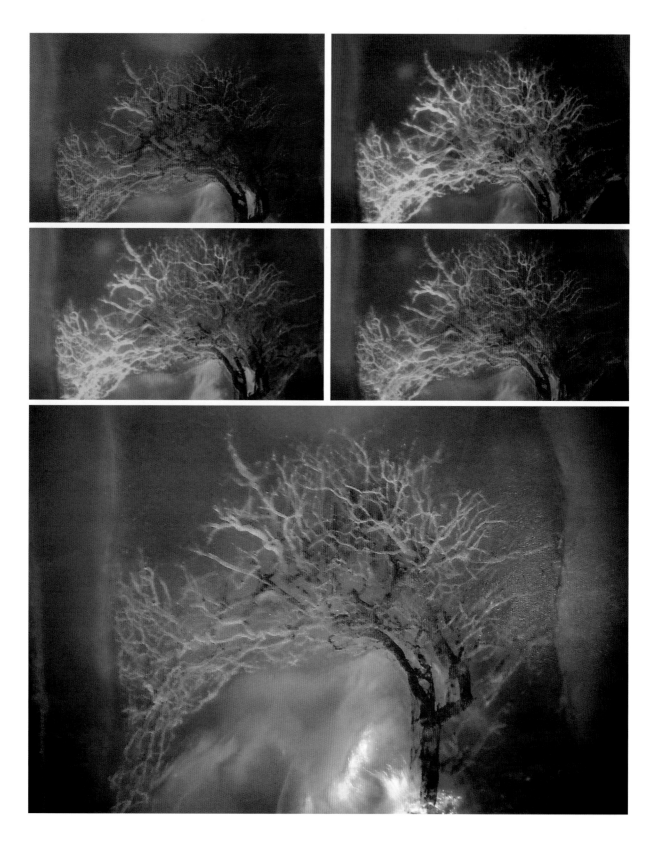

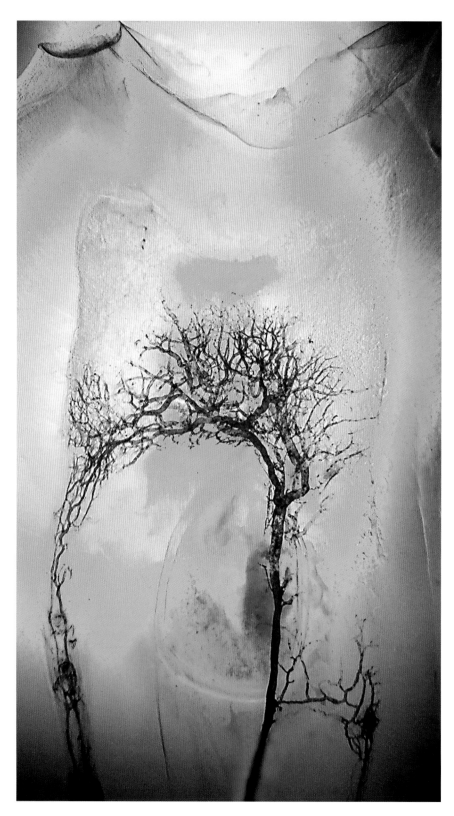

FIG 2-24 A maxillary molar.

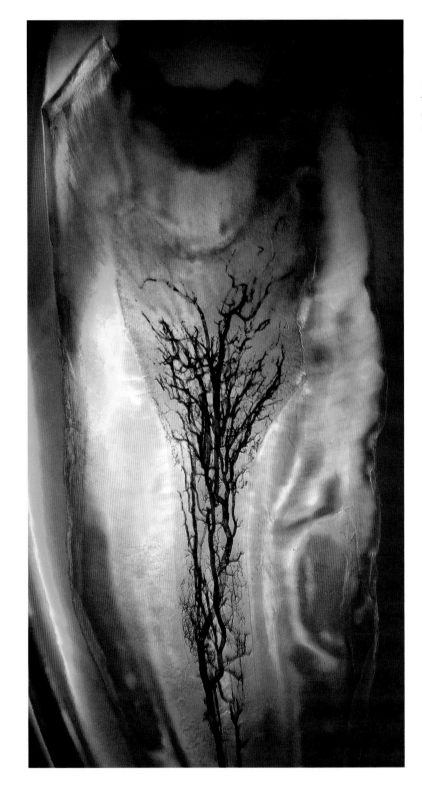

FIG 2-25 One of the most beautiful, complete, and pristine results to date.

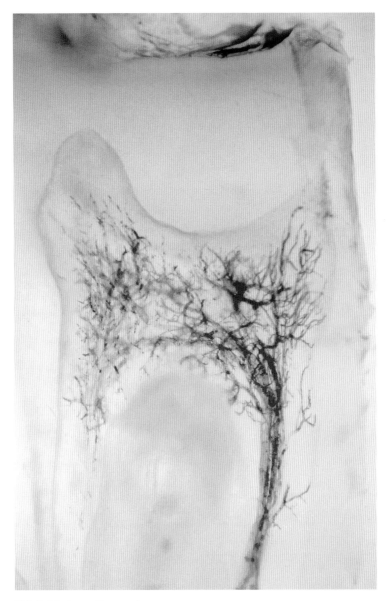

FIG 2-26 A maxillary molar.

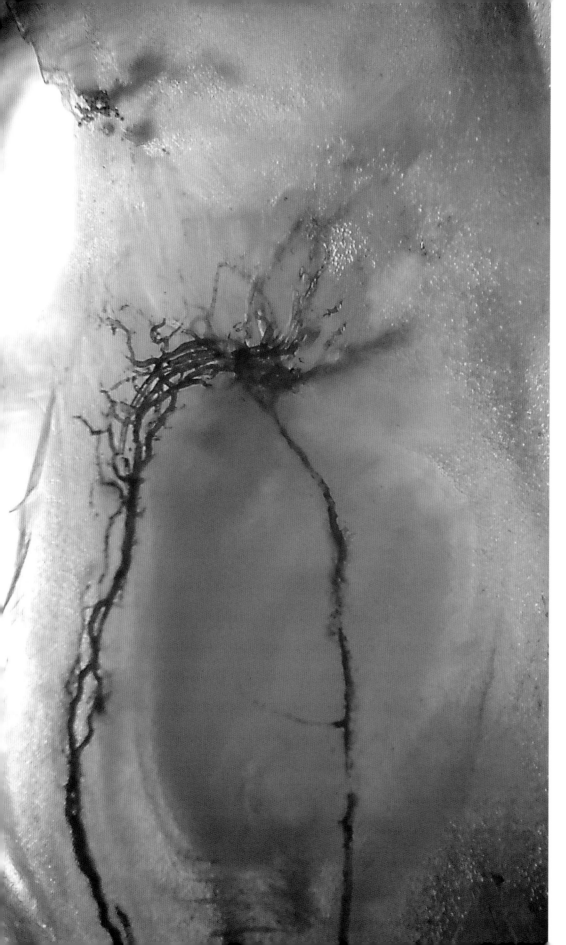

FIG 2-27 Focused on the channel anatomy leading to the pulp chamber.

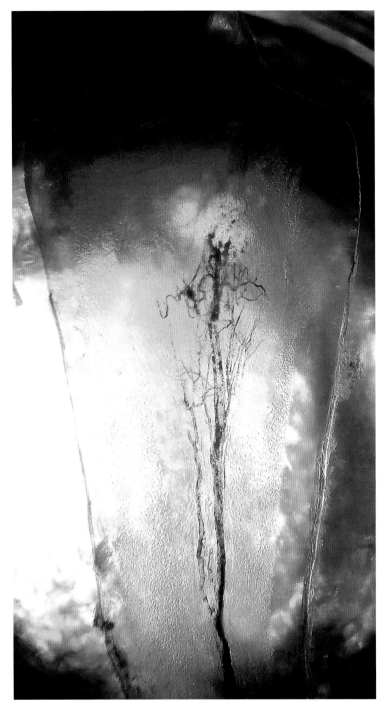

FIG 2-28 A maxillary premolar.

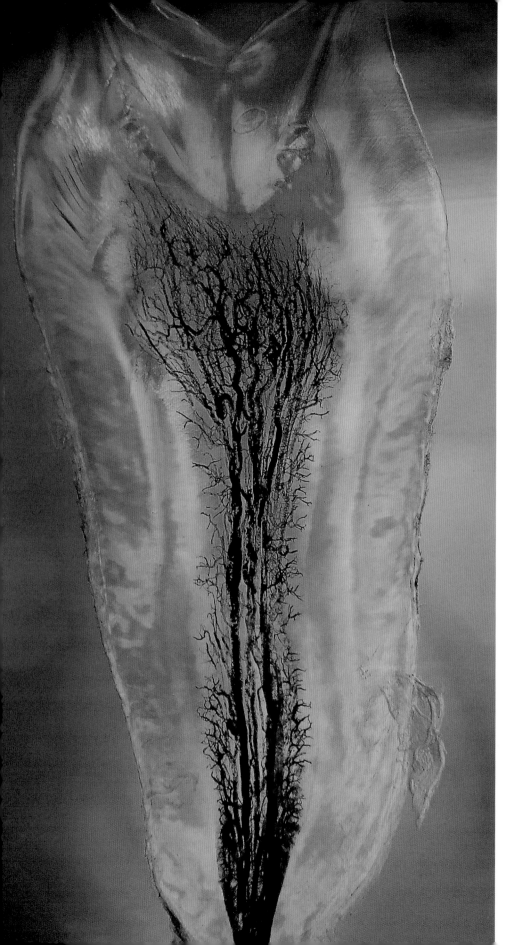

FIG 2-29 Another maxillary premolar.

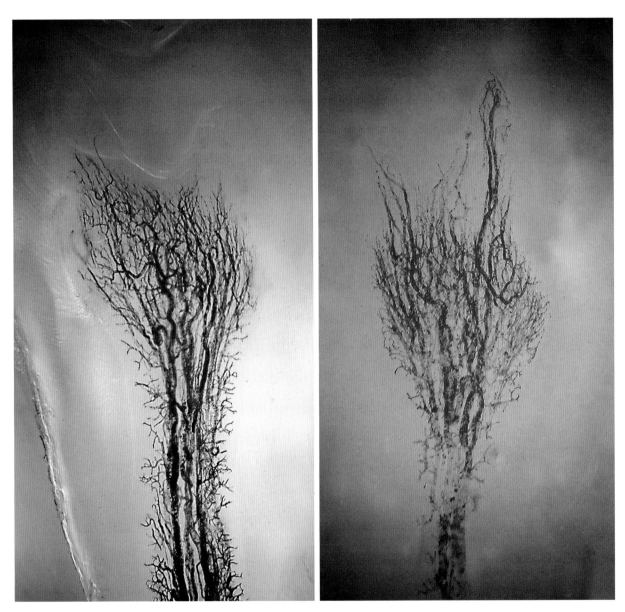

FIG 2-30 Two examples of microlooping
of vascularity in the pulp horn.

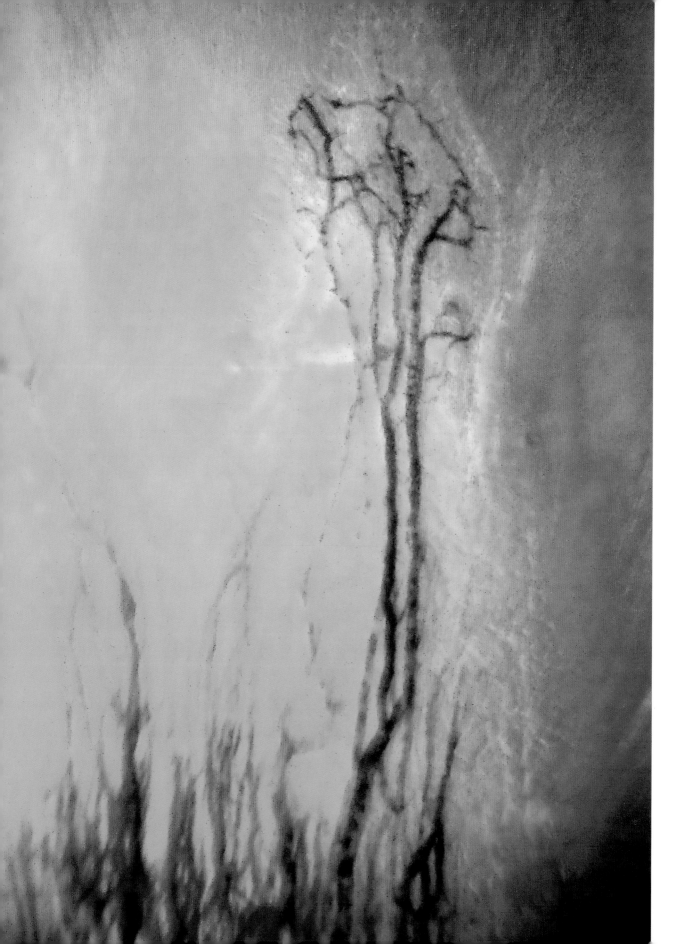

FIG 2-31 (left) Arteriovenous anastomosis in a pulp horn.

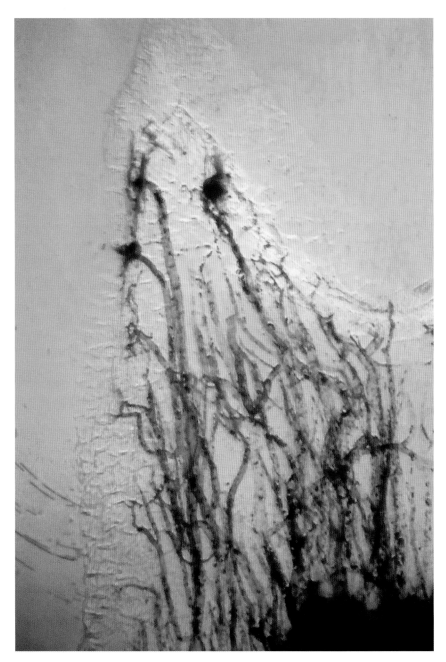

FIG 2-32 Dentin-odontoblastic-vascular interface.

FIG 2-33 Gross view of the vascularity of a maxillary molar.

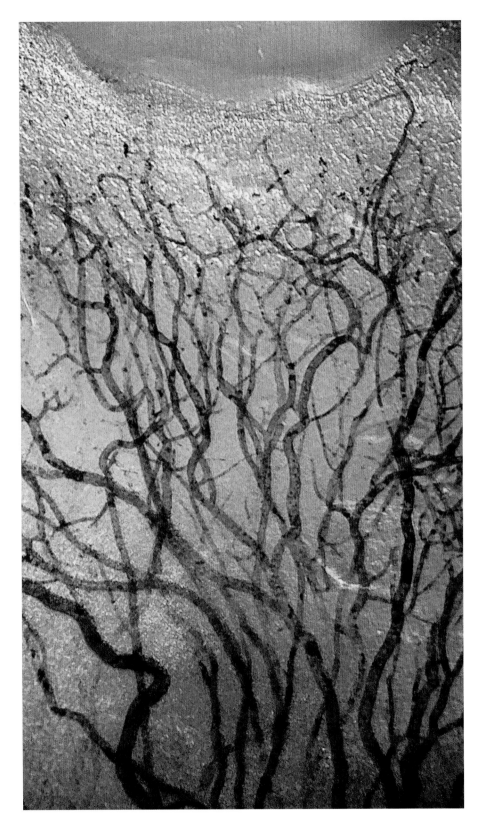

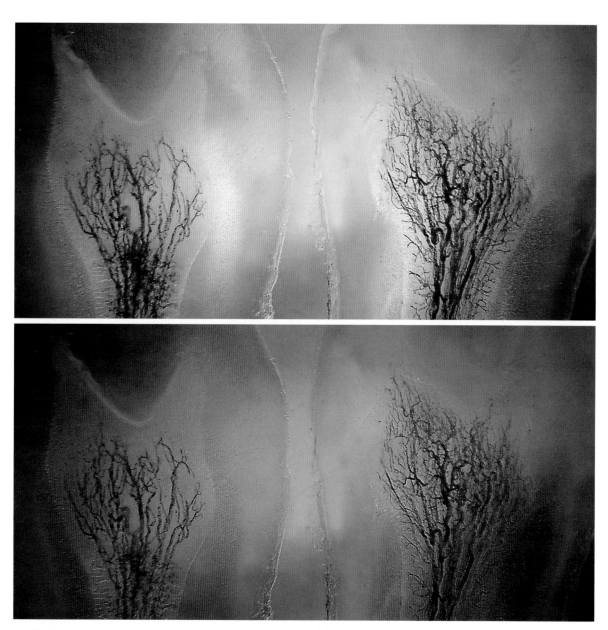

FIG 2-34 A comparative anatomical look at the exposed vascularity of maxillary first premolars.

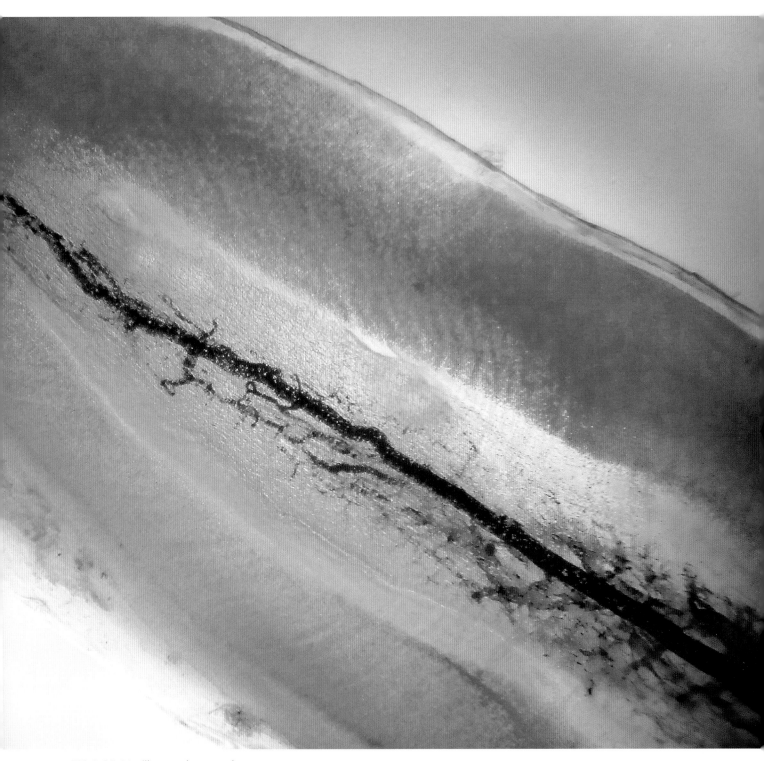

FIG 2-35 Maxillary canine vascular exposure.

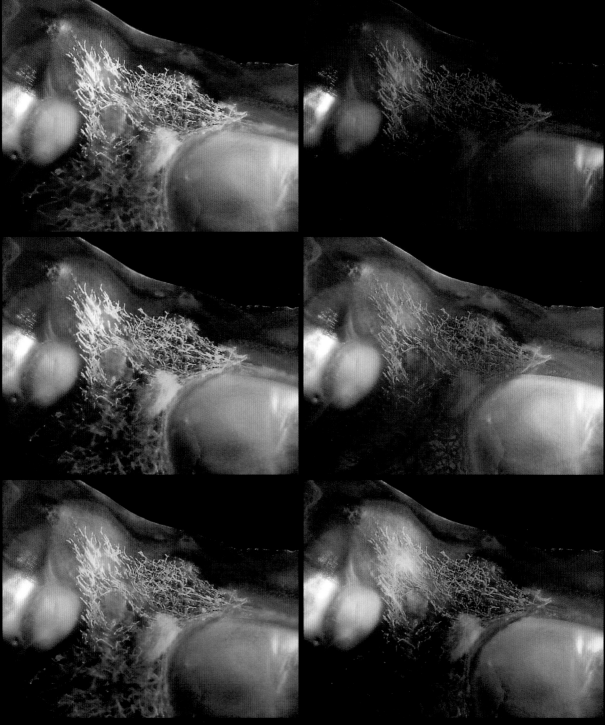

FIG 2-36 A maxillary molar.

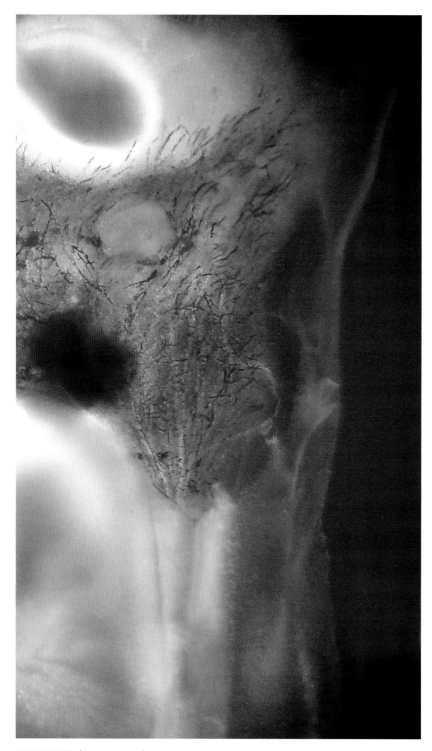

FIG 2-37 Pulp stones and nerves are present.
A hint of the nerve plexus can be visualized.

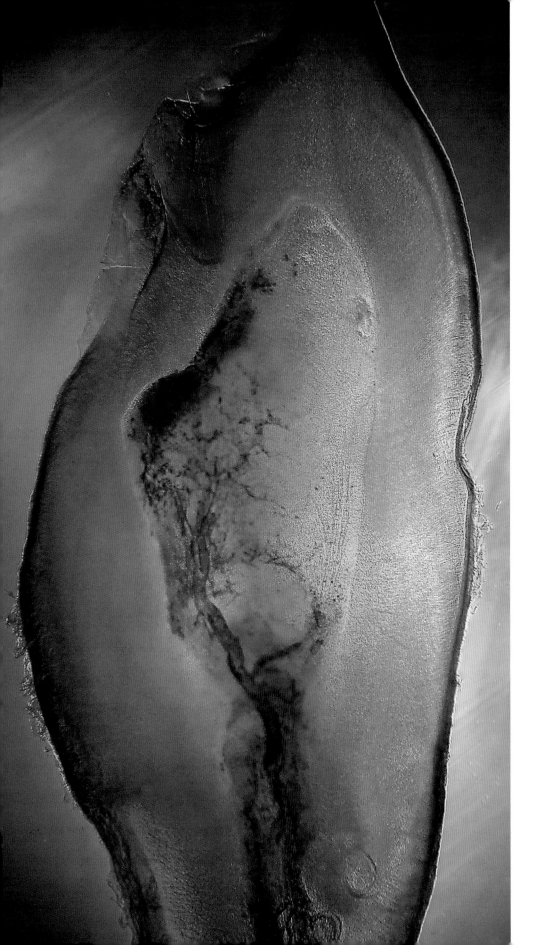

FIG 2-38 A histologic look at late-stage irreversible pulpitis. Pulp stones and Raschkow plexus are seen.

FIG 2-39 Distal channel with multiple small denticles.

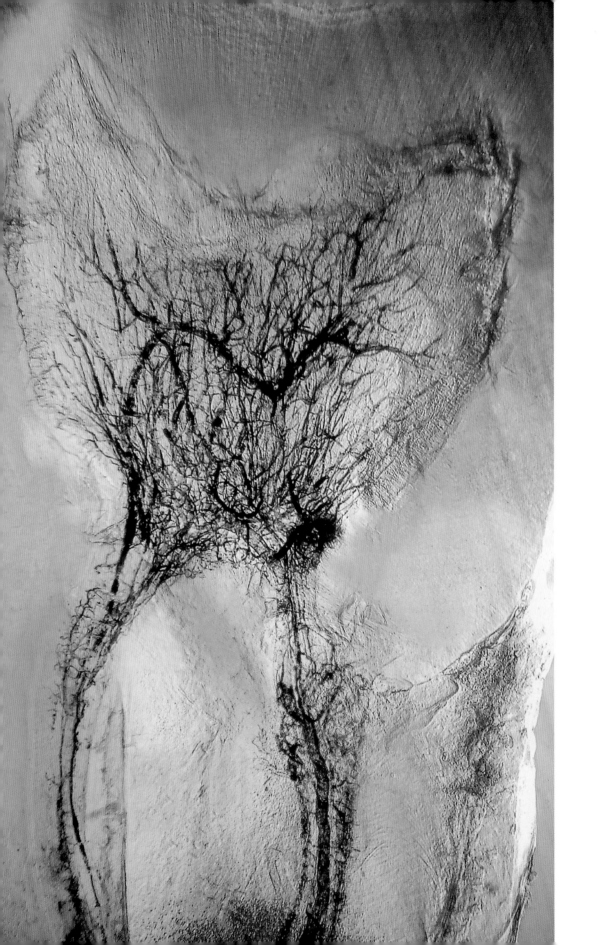

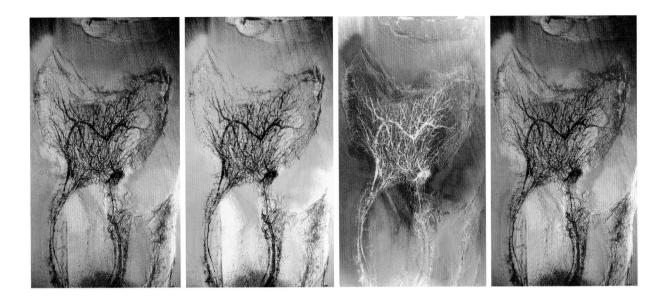

FIG 2-40 Maxillary molar. A definitive look at arterial and venous structures. The venous flow is the large vessel seen in the chamber. The structure moves from right to left and down the left side channel orifice.

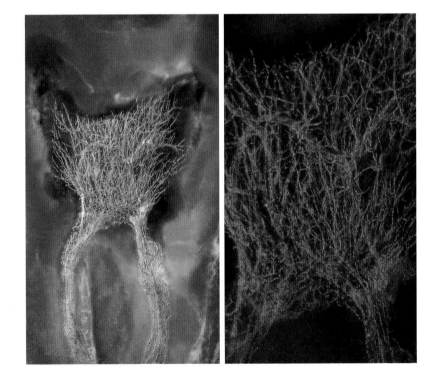

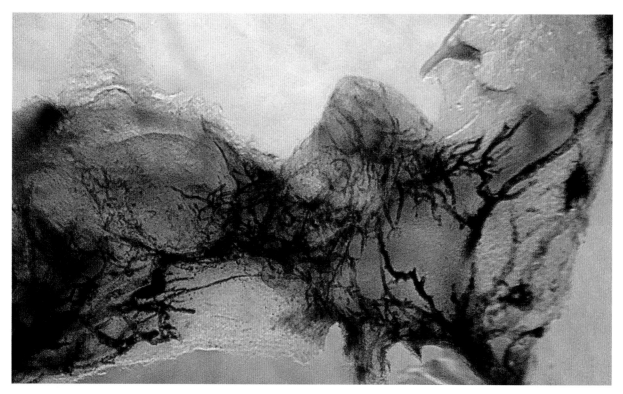

FIG 2-41 A look at early necrosis.
A large pulp stone is seen.

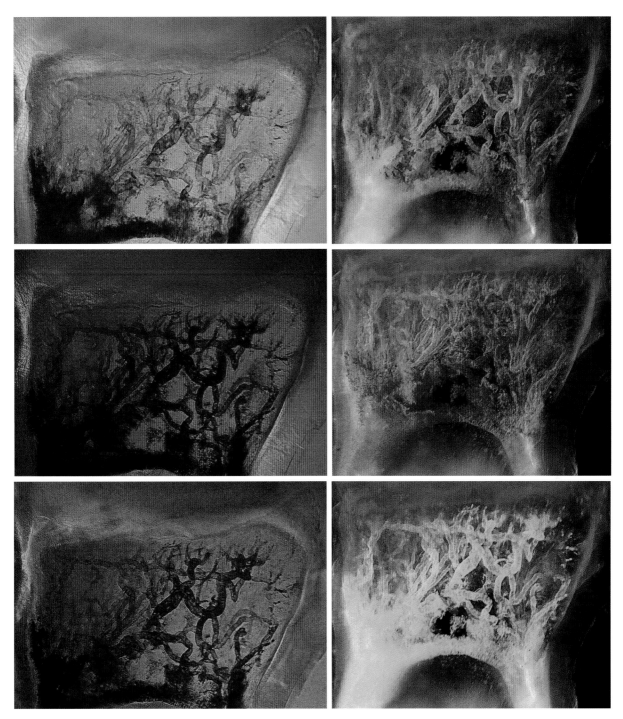

FIG 2-42 An aberrant-looking chamber that potentially should be in the pathology chapter. This represents what could be a lifelong research project of discovery. This could be related to genetics, a drug interaction, or a representation of a diagnosable disease or an undiagnosed disease that manifests itself in teeth.

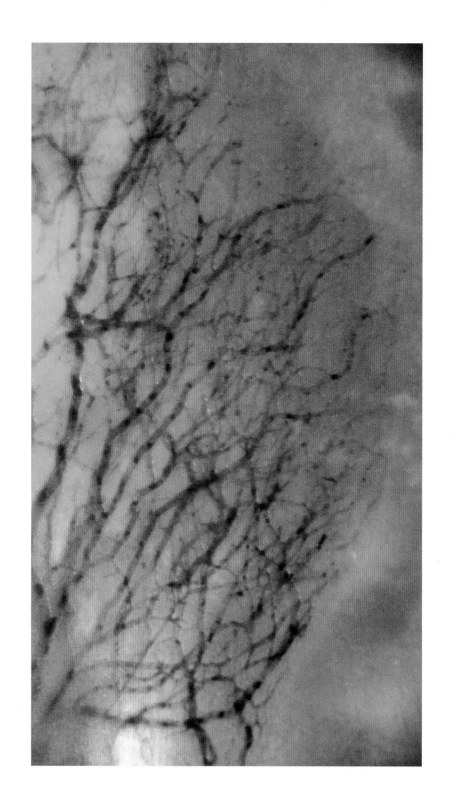

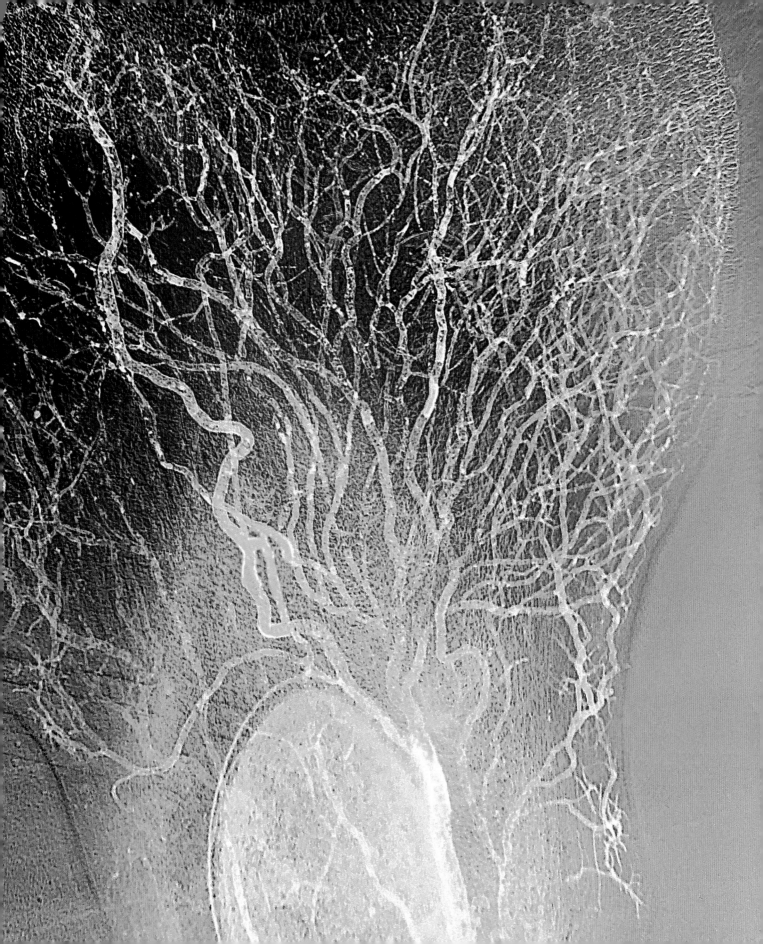

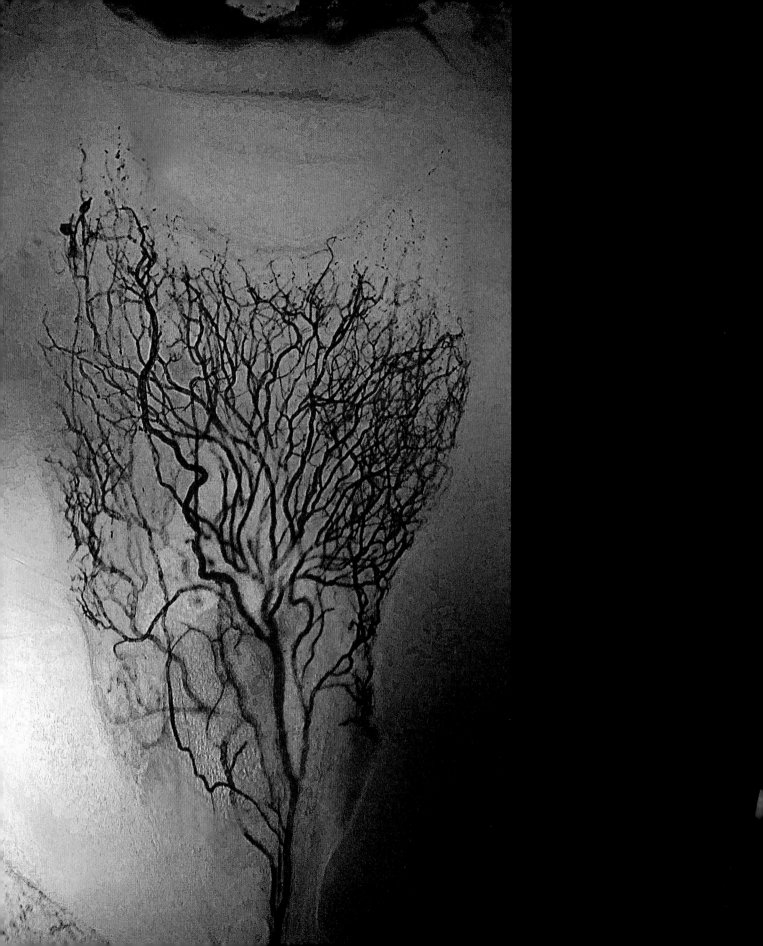

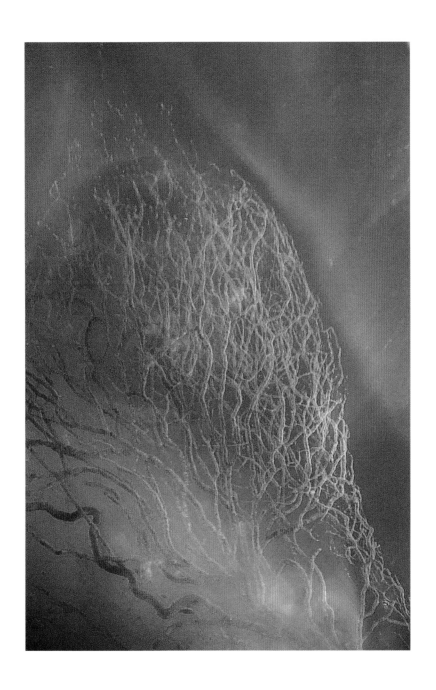

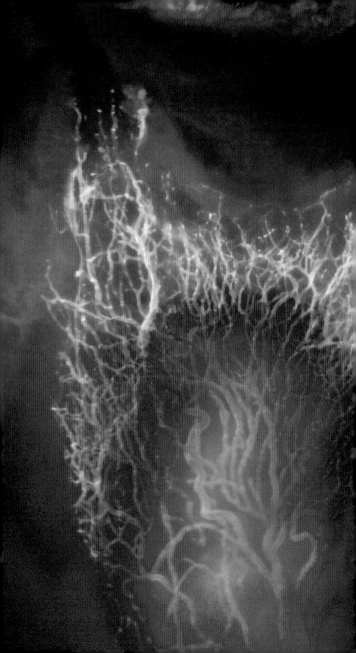

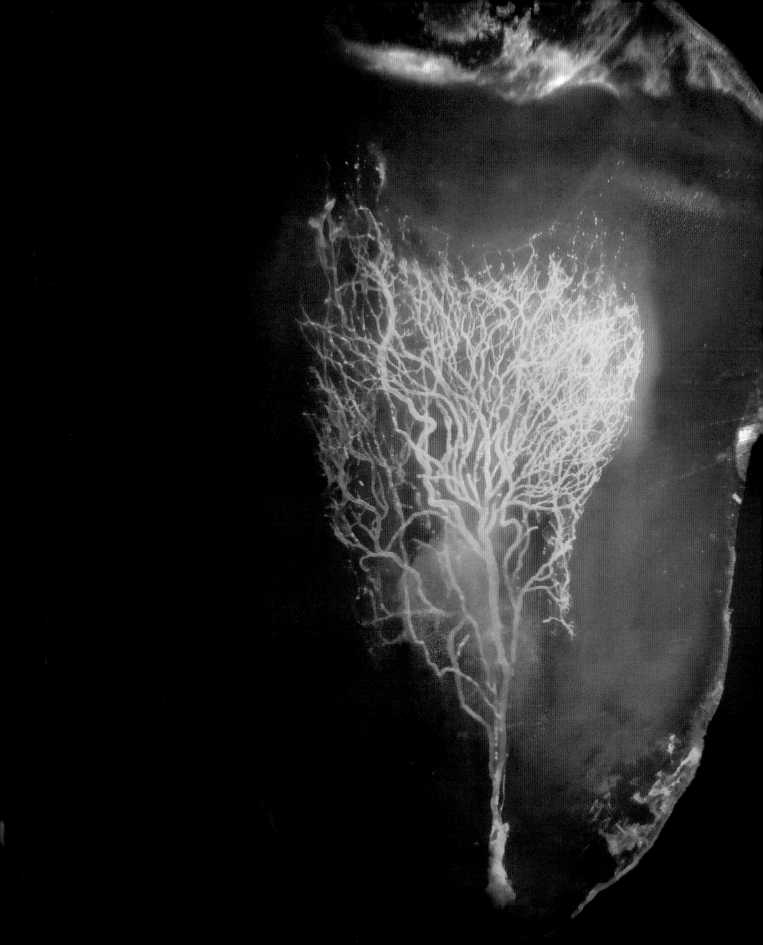

3

Apical Anatomy

The neurovascular bundle entering a tooth is mythic, and its complexity is not well understood. This is in part because it has not been visualized routinely. The images in this chapter are a relatively small sampling that still bring up important questions. Where does the nerve enter or exit? What is the purpose of this varied presentation of anatomy? I am proud to be able to reveal any sort of vascularity in this minute and complex region, even if it only leads to more questions. The only conclusion at this time is that most lateral channels contain a blood vessel of some sort.

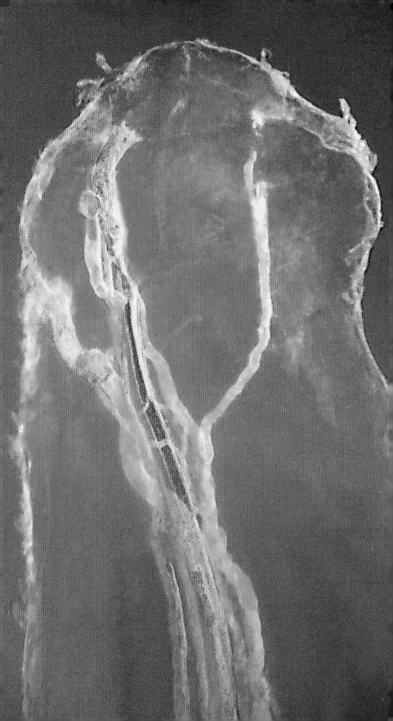

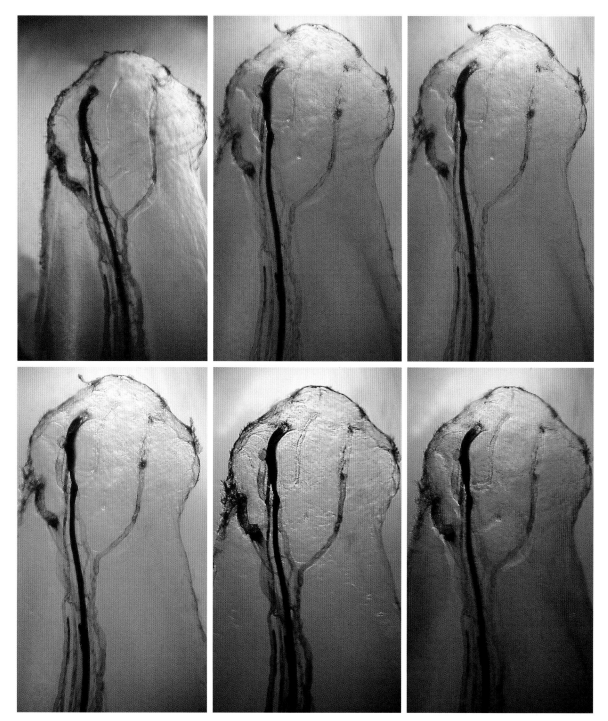

FIG 3-2 What has been considered to be lateral channels in the past are actually vascular channels. The channels without contrast stain are considered to contain innervation.

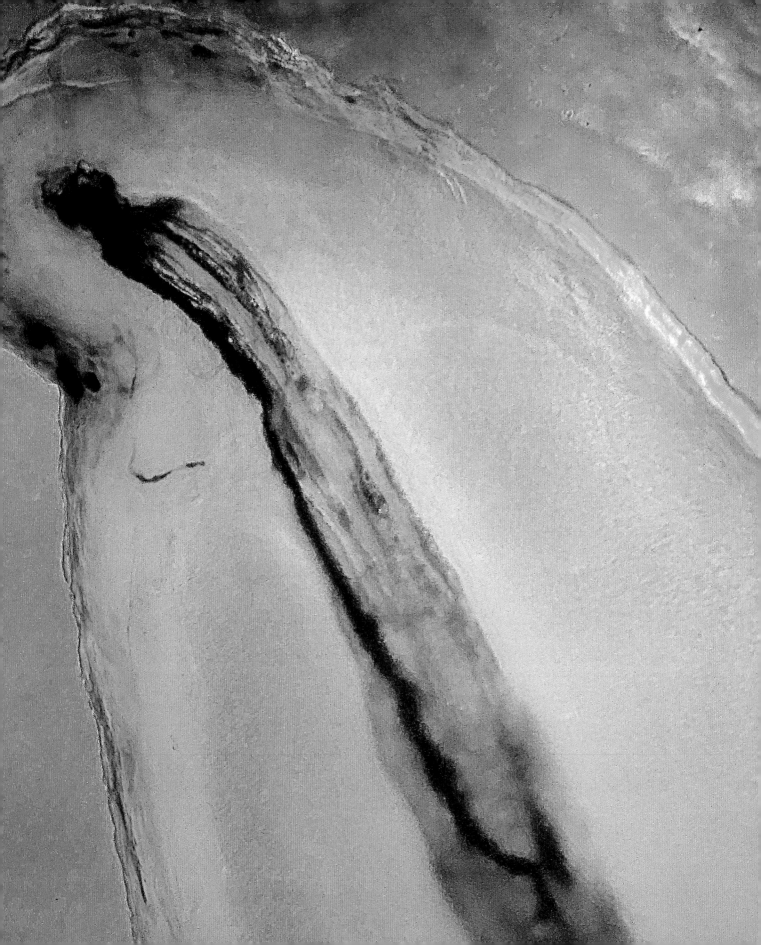

FIG 3-3 (left) An early result demonstrating a "constriction." An independent channel is observed.

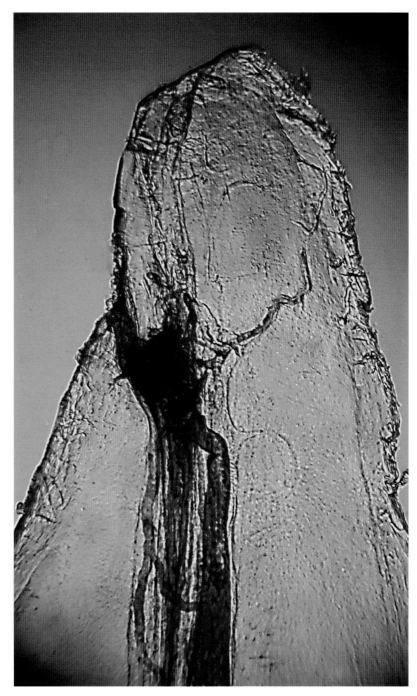

FIG 3-4 There is a nice apical recurrent loop to the right of the demonstrated apical systems.

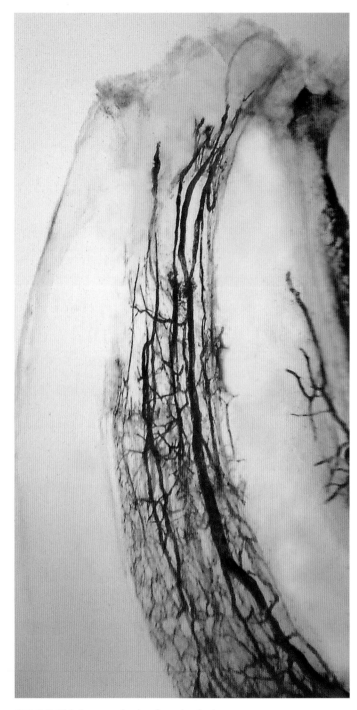

FIG 3-5 This is an underdeveloped apical region, but it is developed enough to have formed vascularization.

In anatomy, a *canal* refers to a tubular passage or channel that connects different regions of the body. There is obvious ambiguity in this definition, which leaves open debate as to what is the correct term to be used when referring to anatomical passageways in human teeth. A deeper dive into Mirriam-Webster reveals a nuance in usage that specifies a *canal* as being an artificial or man-made passageway, whereas a *channel* is used continually to denote a "natural" or tubular enclosed passage. When all the available definitions of a *channel* versus *canal* are taken into consideration, it is evident that *canal* is associated with the terms *man-made* or *artificial*. By contrast, in all of the available contexts, a *channel* is considered a *natural* pathway. Therefore, I would propose that the systems and passageways within the human tooth are *channels* until they are shaped and affected by the practitioner; it is that preparation that forms them into *canals*.

In this chapter, you will see that channel vascular anatomy is unique in its presentation, when compared to the apical and chamber formations. The resulting visuals literally defy logic or the laws of gravity or physics. The distributions go against any sort of "law of conservation of matter." The photographs of this series of results focus on interesting findings of blood vessels as they travel to and from the pulp chamber and apical region. (In the laboratory, I have been able to recreate the pseudo function of these vessels, and what takes place in terms of fluid mechanics is beyond amazing.)

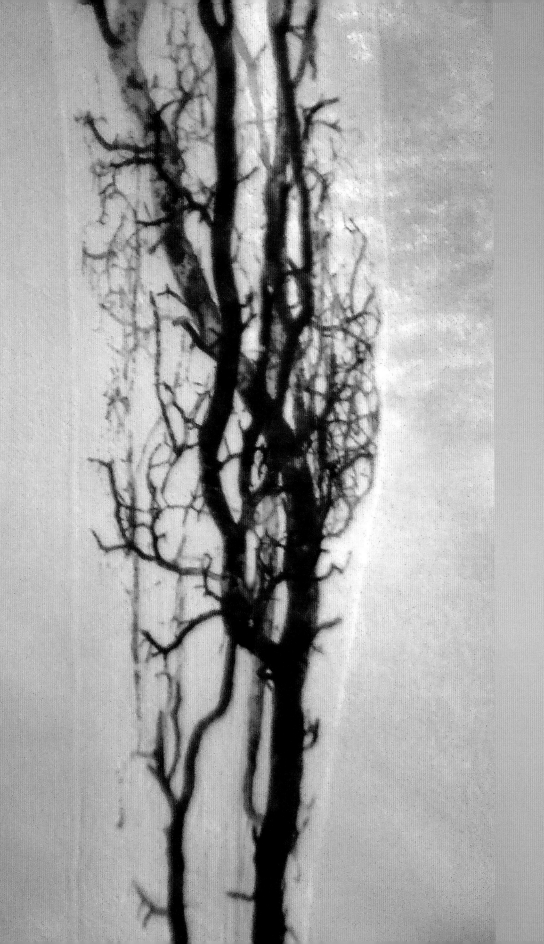

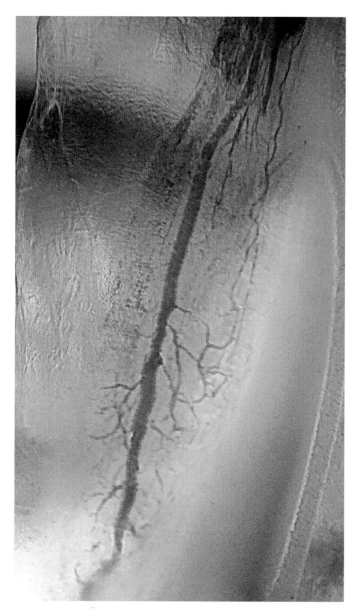

FIG 4-1 The volume and complexity of the vascular distribution within human teeth truly has no current explanation.

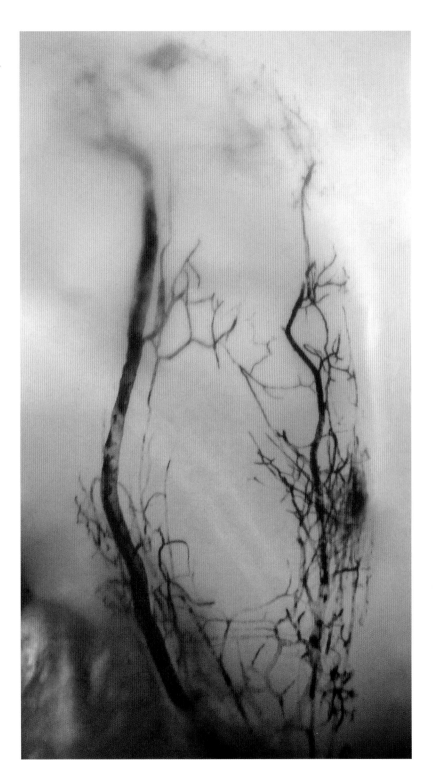

FIG 4-2 What would be classified as "MB1/MB2" to the uneducated pulp channel anatomists in a maxillary first molar.

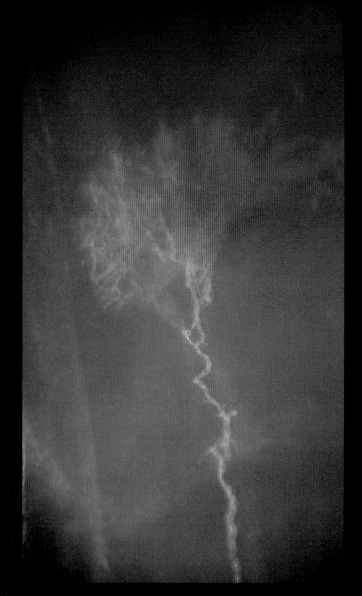

FIG 4-3 Vasculature traveling through the distal channel of a mandibular molar interfacing with the pulp chamber vascular bed.

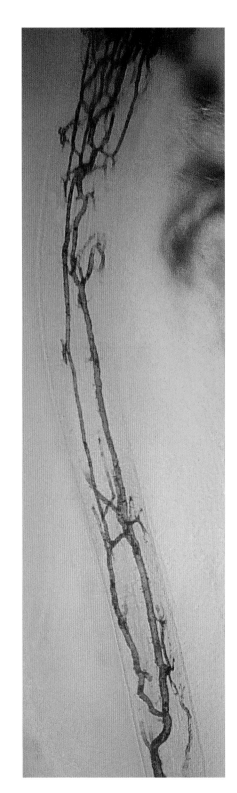

FIG 4-4 Vascularity in a palatal channel of a maxillary molar.

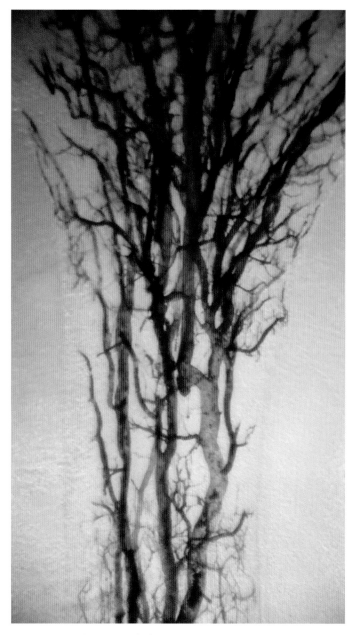

FIG 4-5 Vasculature interfacing with the inside channel wall.

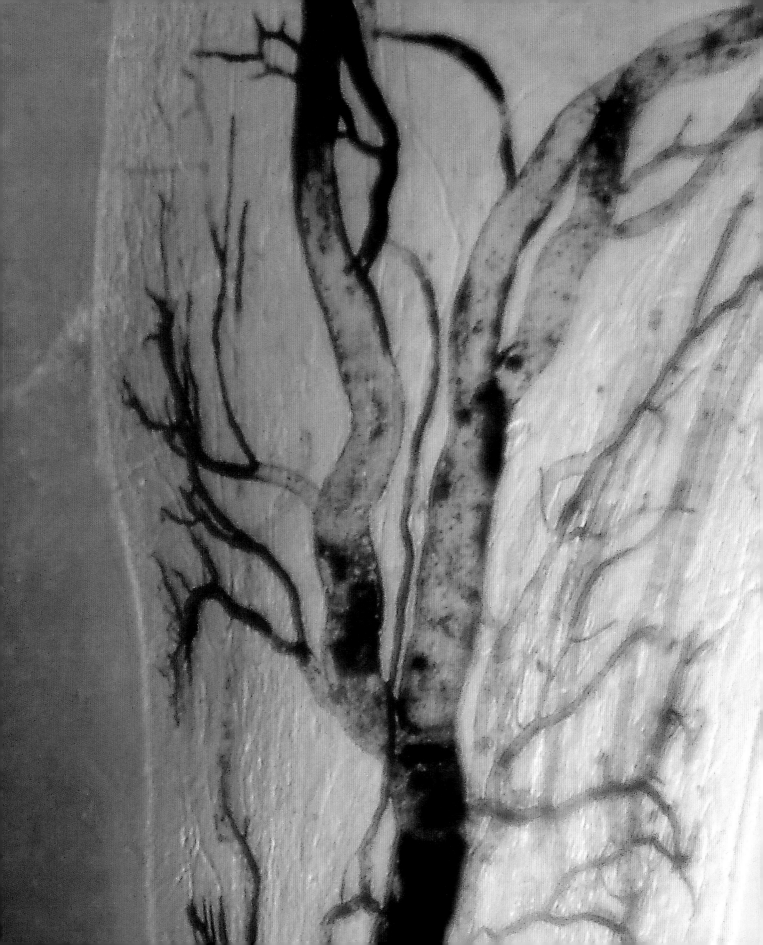

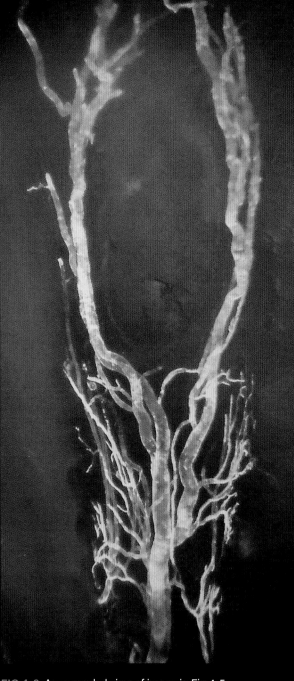

FIG 4-6 An expanded view of image in Fig 4-5.

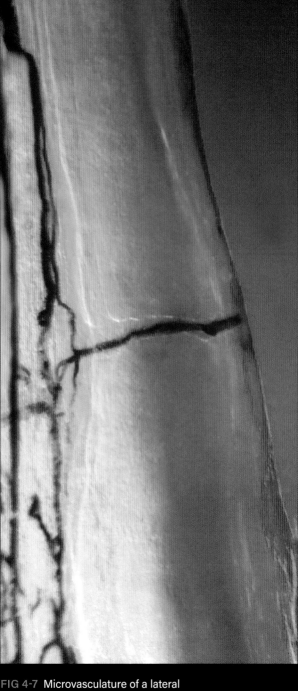

FIG 4-7 Microvasculature of a lateral channel. Channel is to the *left* and the external root surface is to the *right*.

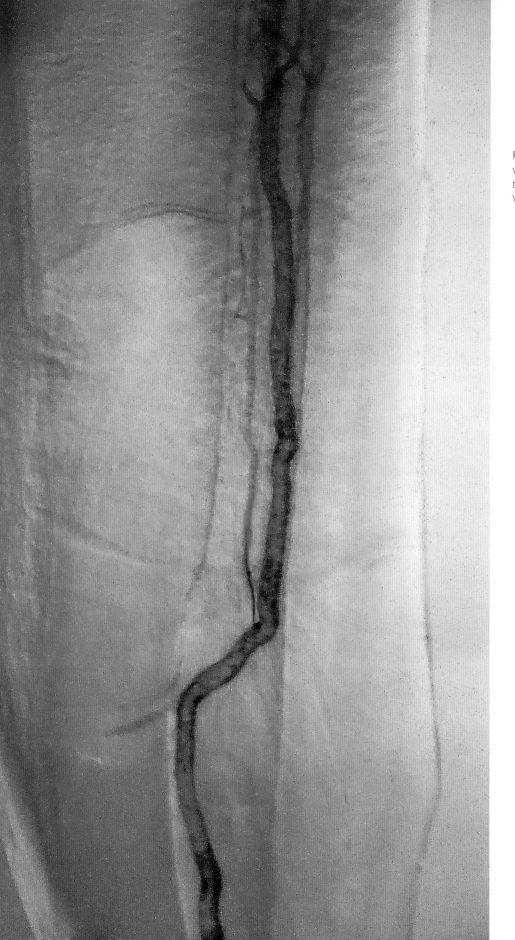

FIG 4-8 Why the shift of the vessel from left to right only to give off another branch? Vascularization defies gravity.

5

Dual Staining

In this group of specimens, I was able to stain the venous flow in one color and the arterial flow in another color. For ease and to keep with tradition, I then used Photoshop to make the venous flow *blue* and the arterial flow *red*, unless otherwise noted. Therefore, the depicted colors are not the actual stain colors used in the lab.

With all the considerations given to simply exposing the vascularity itself, it is a further miraculous and phenomenal accomplishment to be able to present the vascular distribution where two stains of different colors were introduced into separate supply sources of the tooth.

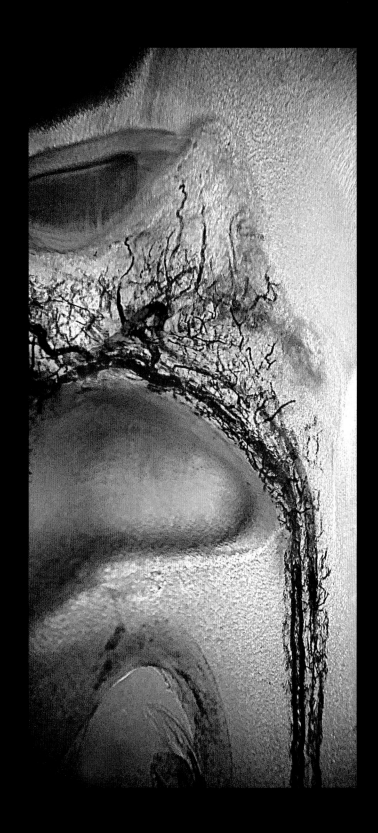

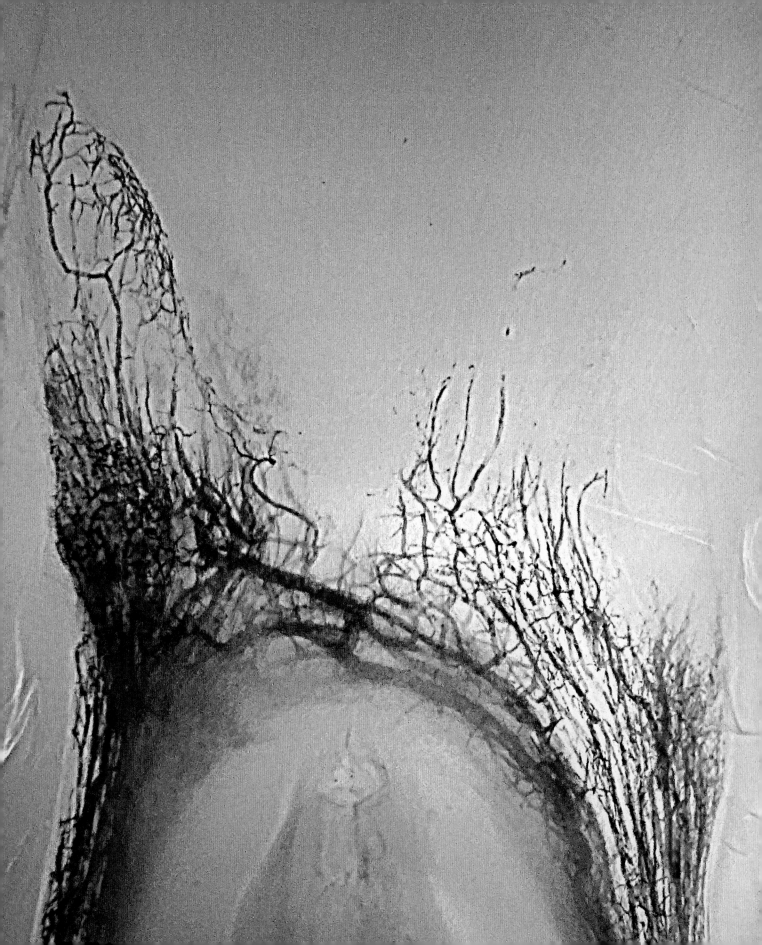

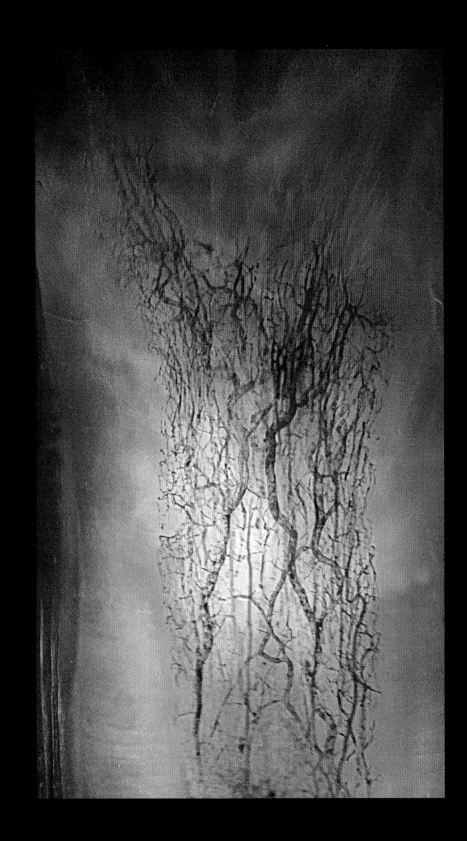

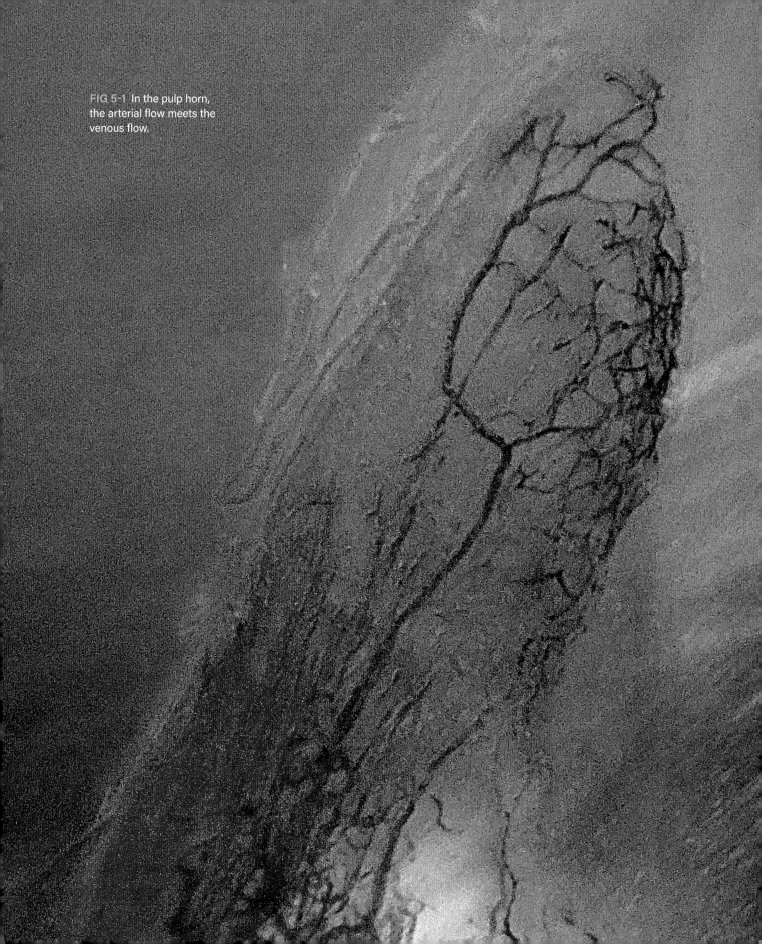

FIG 5-1 In the pulp horn, the arterial flow meets the venous flow.

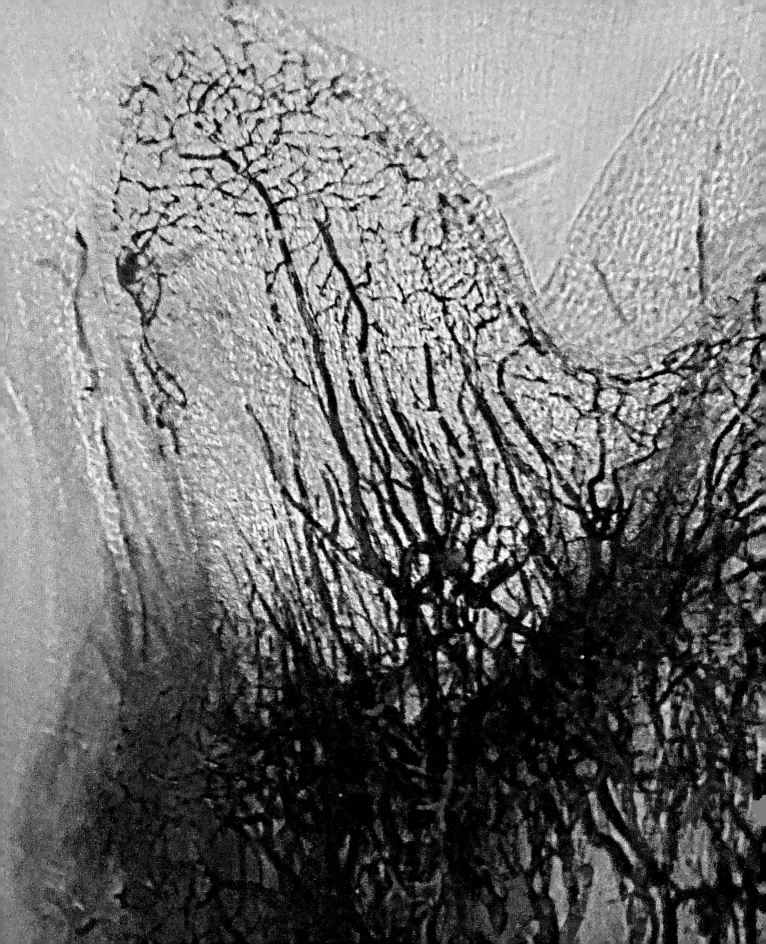

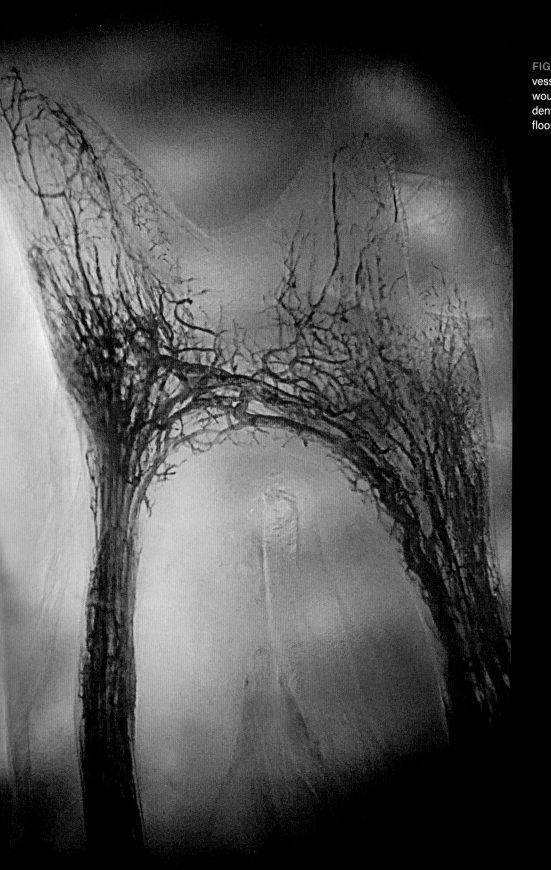

FIG 5-2 Note the large central vessel in the chamber. This would be associated with the dentin map found on the pulp floor of a tooth chamber.

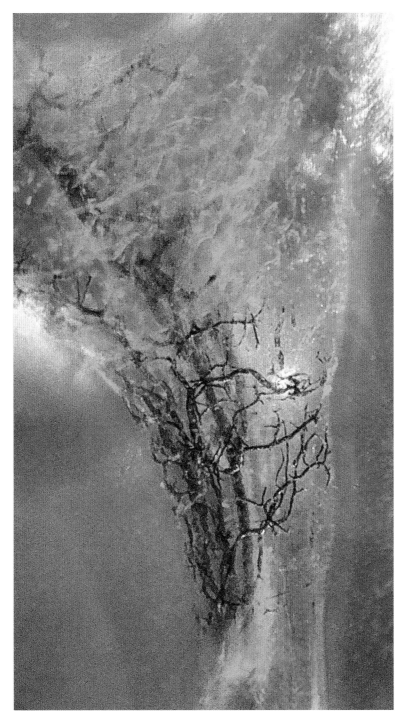

FIG 5-3 Arterial flow is in *black*.

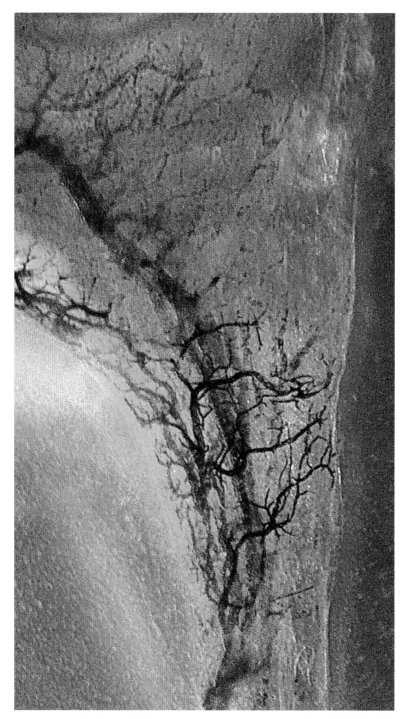

FIG 5-4 Arterial flow is in *black*.
Venous flow is in *red*.

This chapter represents some of the early exposures of vascularity of internal structures of a human tooth. These are results of methods not represented in the other chapters of this book. As I was running experiments to expose structures of the internal anatomy of human teeth, it was these accidental occurrences in my laboratory that led to the methods and results featured in the rest of this book.

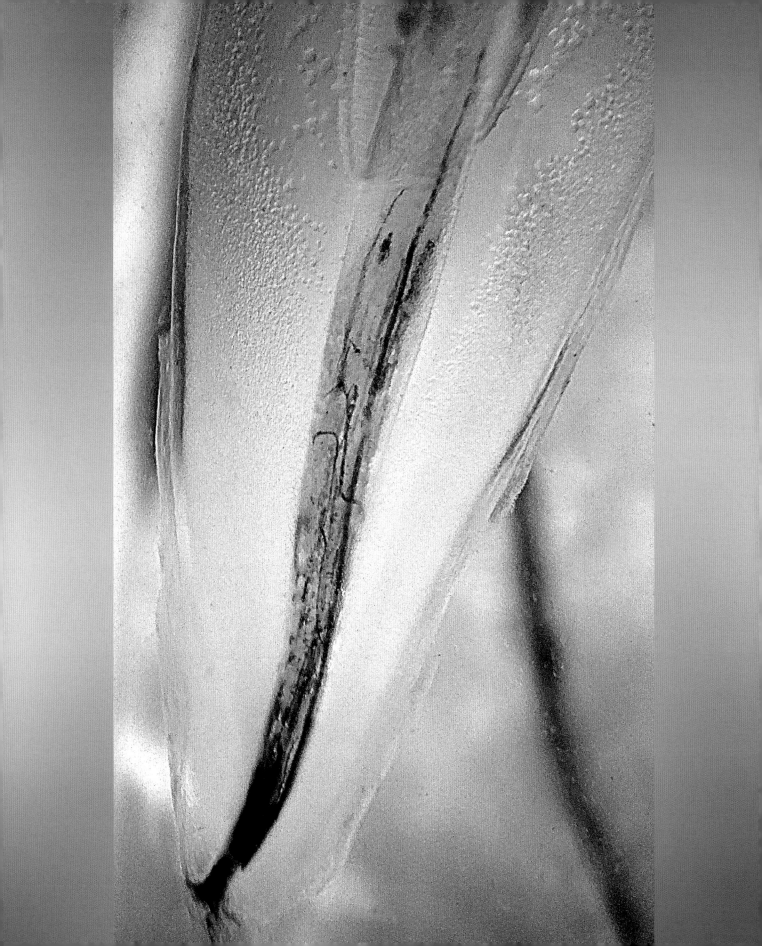

FIG 6-1 Extracanal invasive resorption in a maxillary central incisor.

FIG 6-2 Close-up of the incisal edge of the tooth.

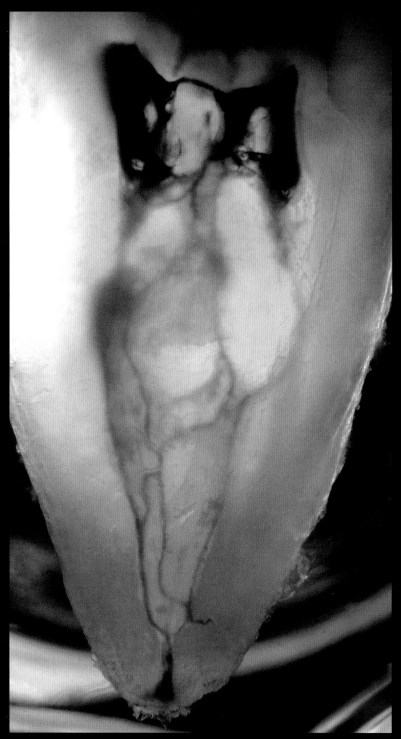

FIG 6-3 Mesial root of a mandibular right first molar.

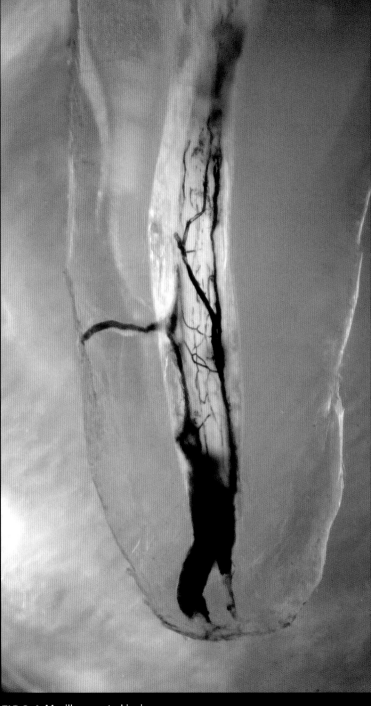

FIG 6-4 Maxillary central incisor.

FIG 6-5 Midroot vascularity
in a maxillary central incisor.

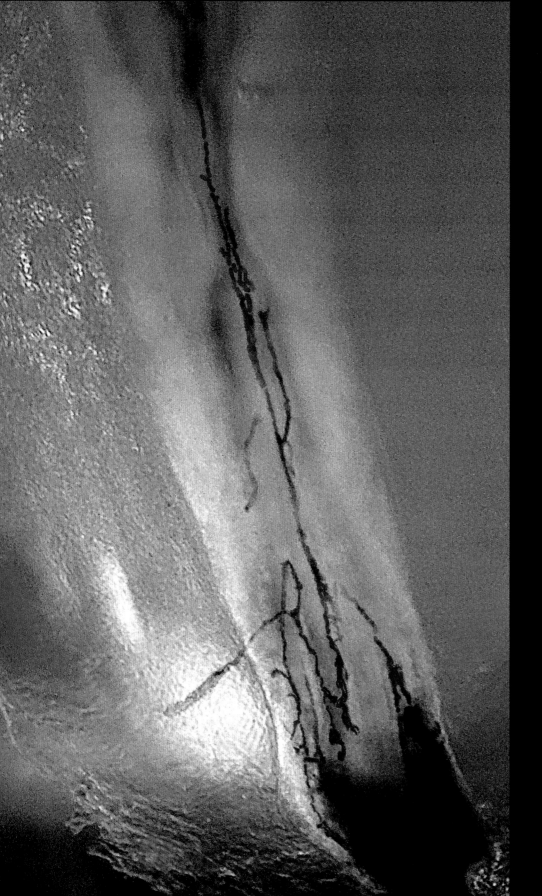

FIG 6-6 Internal looping and an associated lateral channel.

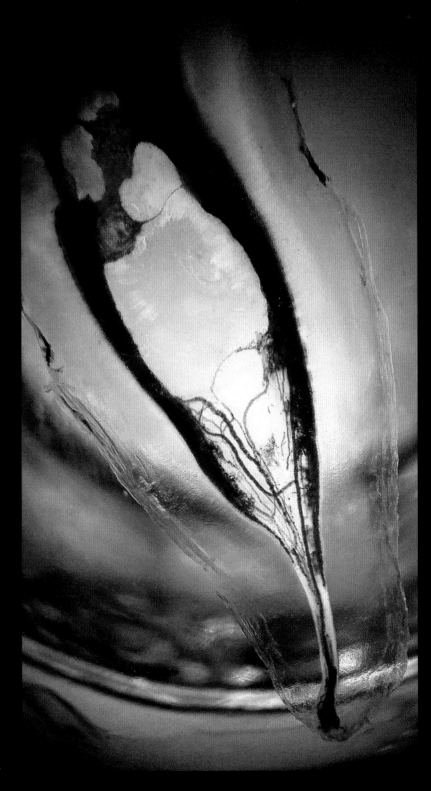

FIG 6-7 Mesial root of a mandibular second molar.

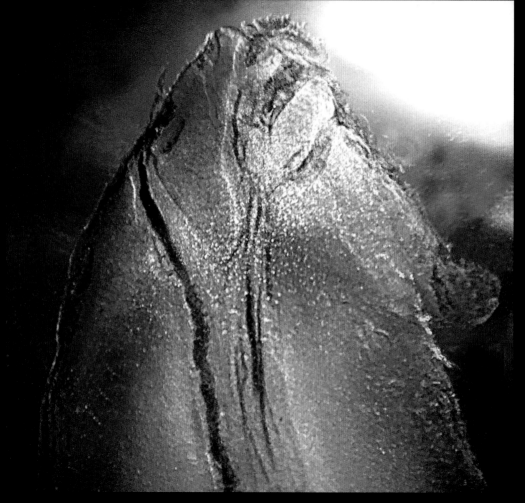

FIG 6-8 Apical region of a

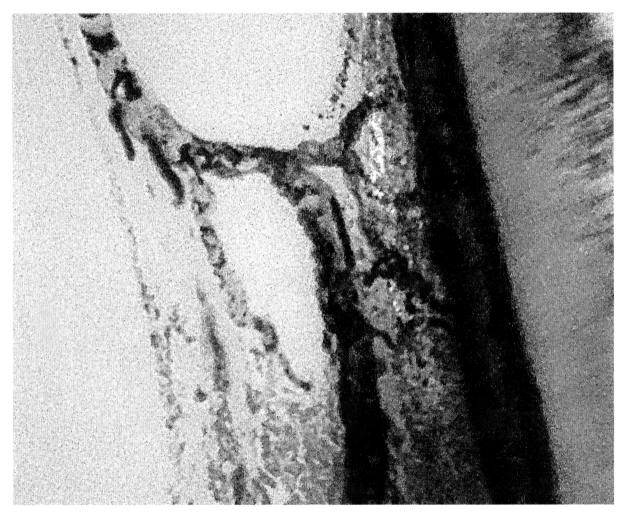

FIG 6-9 Section of a mesial buccal
root of a maxillary first molar.

7

Pathology

Given all the patterns revealed and repeated in chapter 2, I started identifying results that did not make sense or match the majority of other results. Those aberrant outcomes are depicted in this chapter. In addition, all samples were collected such that teeth with known pulpal status were used. Necrotic pulps obviously have obliterated pulps with nothing to expose beside tissue remnants, immune response faculties, and bacteria, but stages of irreversible pulpitis did give some data to visualize what is potentially happening inside a toothache that is still vital.

Finally, there are other variables that can be expected to alter pulpal status, including cancer and drug response from patient medications, but far more data are needed to continue to explore this theme.

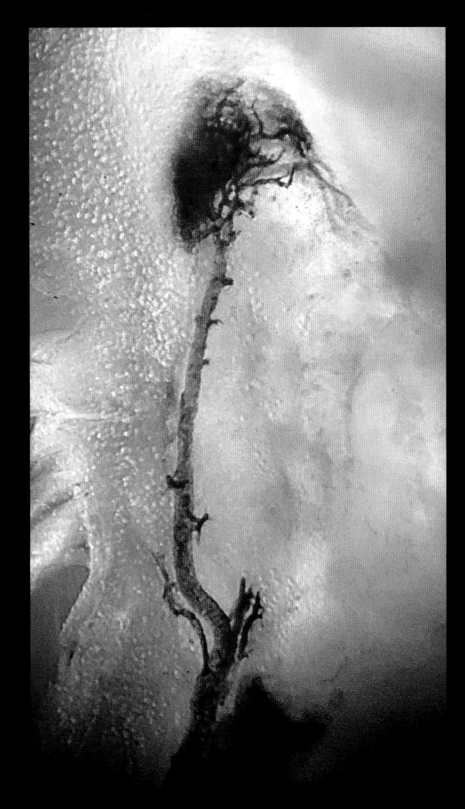

FIG 7-1 Early stage of pulpal necrosis. Note that the branch arms of the major vessel are off.

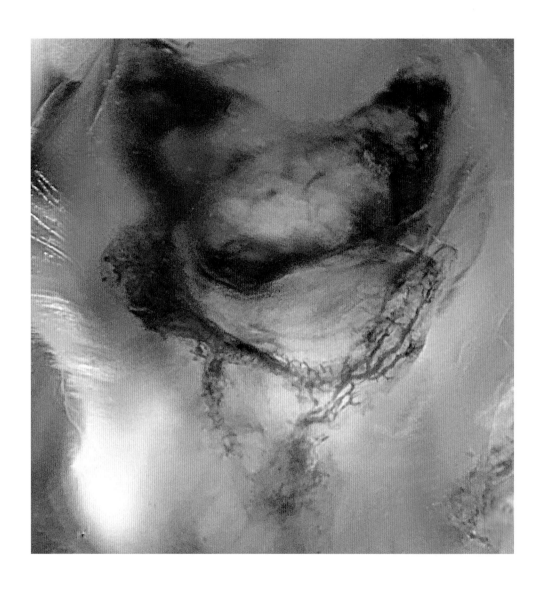

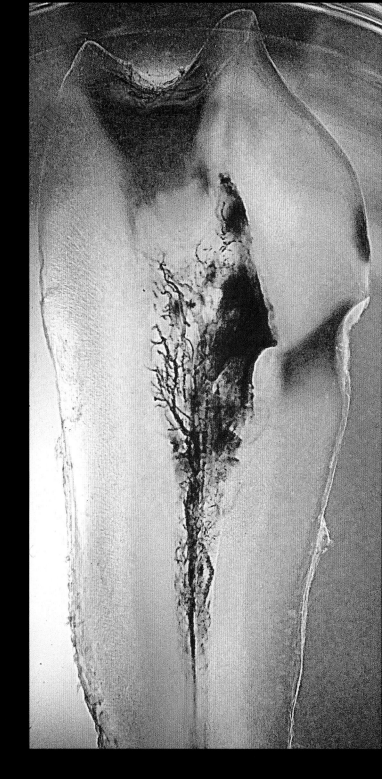

FIG 7-2 Buccal and occlusal decay leading to irreversible pulpitis in a mandibular first premolar.

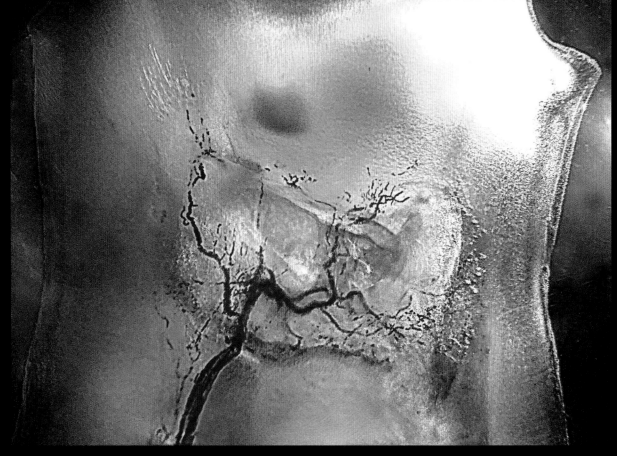

FIG 7-3 A calcified pulp chamber with
associated remaining vascular flow.

FIG 7-4 Extracanal invasive resorption in a mandibular premolar. This is a complex image, and much is to be gained and learned from it. Three contrast mediums were used that did not interact with each other. The vascularity is intact and the pulp fully vital. Multiple invasive tissue points are seen on the left side of the image. The *pink coloration* demarks the extent of the invasion of the tissue. The *yellow* is stained invasive tissue remnants.

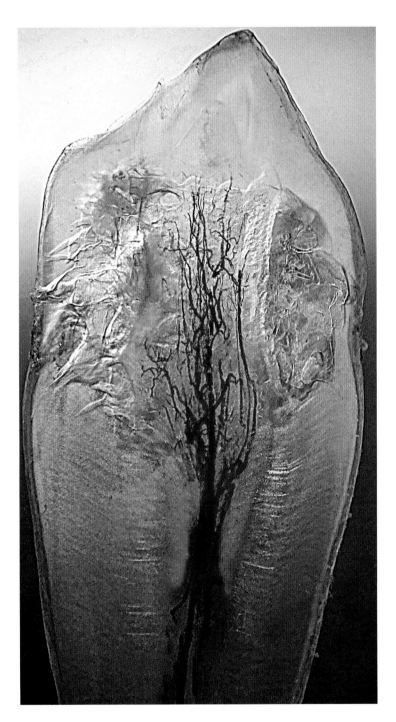

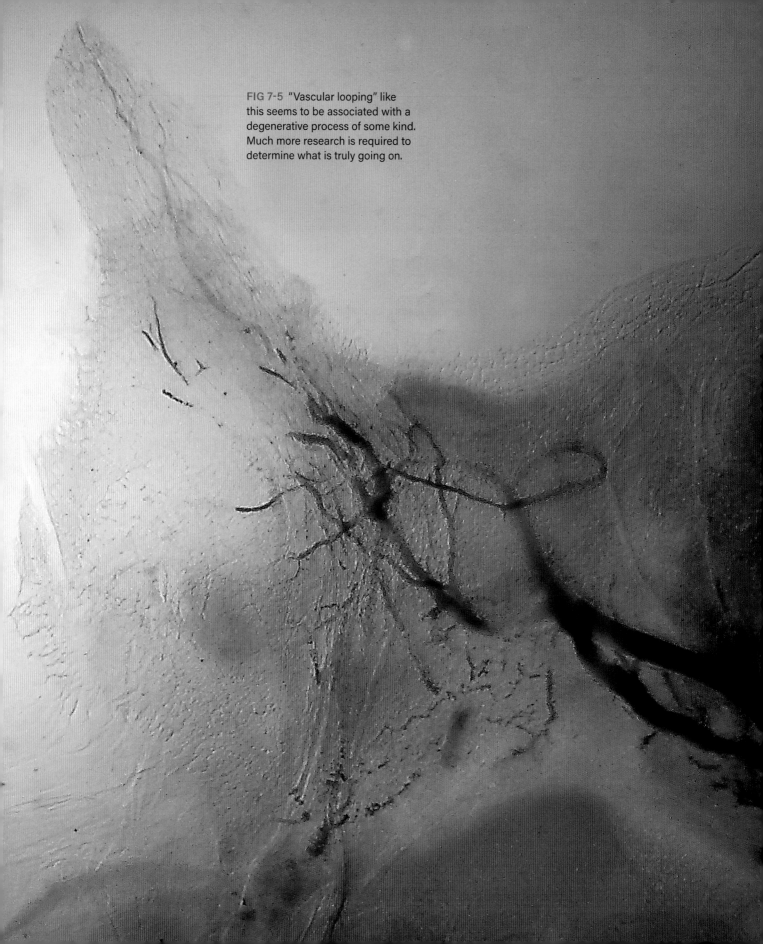

FIG 7-5 "Vascular looping" like
this seems to be associated with a
degenerative process of some kind.
Much more research is required to
determine what is truly going on.

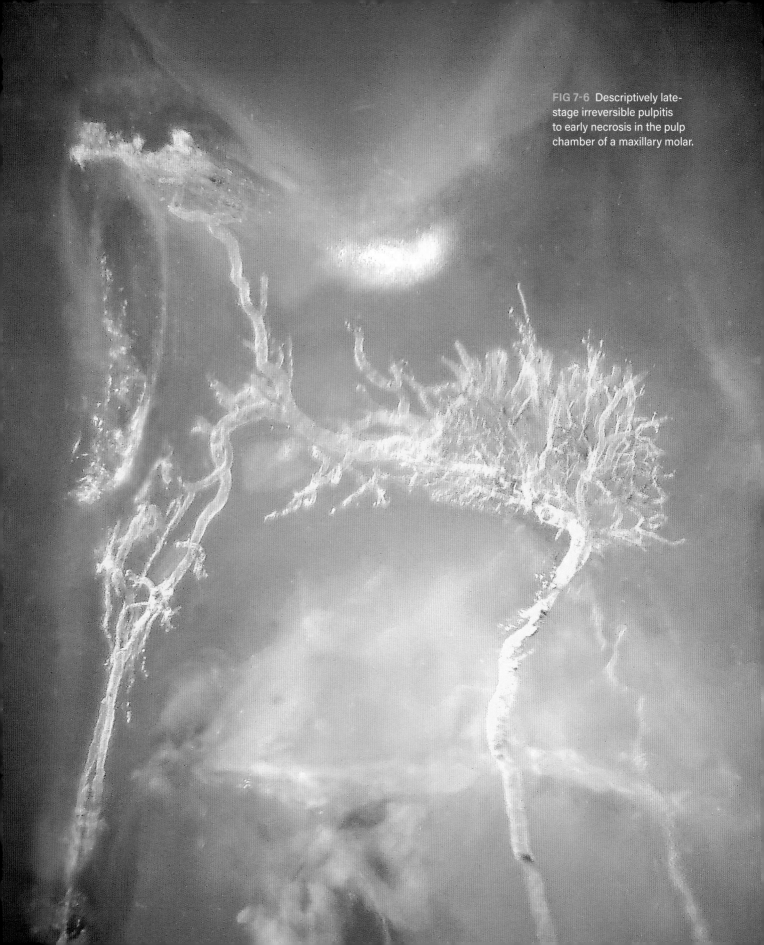

FIG 7-6 Descriptively late-stage irreversible pulpitis to early necrosis in the pulp chamber of a maxillary molar.

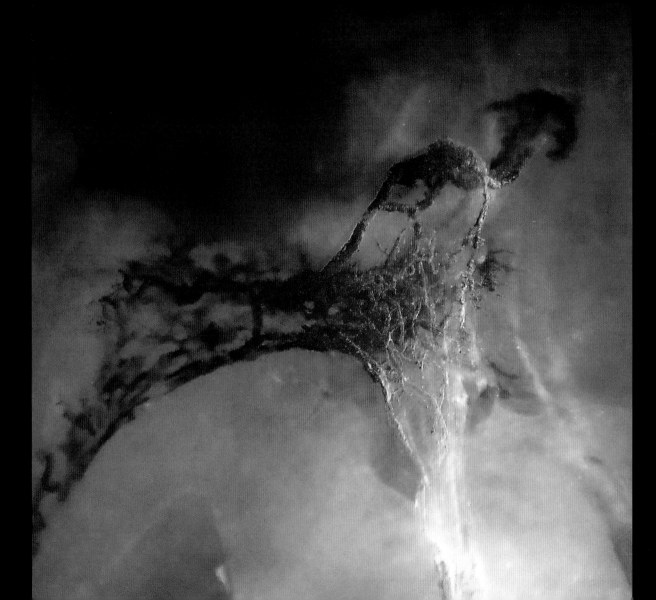

FIG 7-7 Unknown but suspected late-stage irreversible pulpitis to early stage necrosis in the pulp chamber of a maxillary molar.

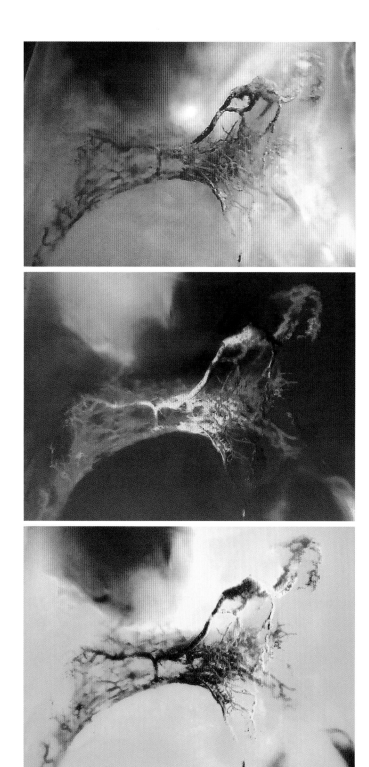

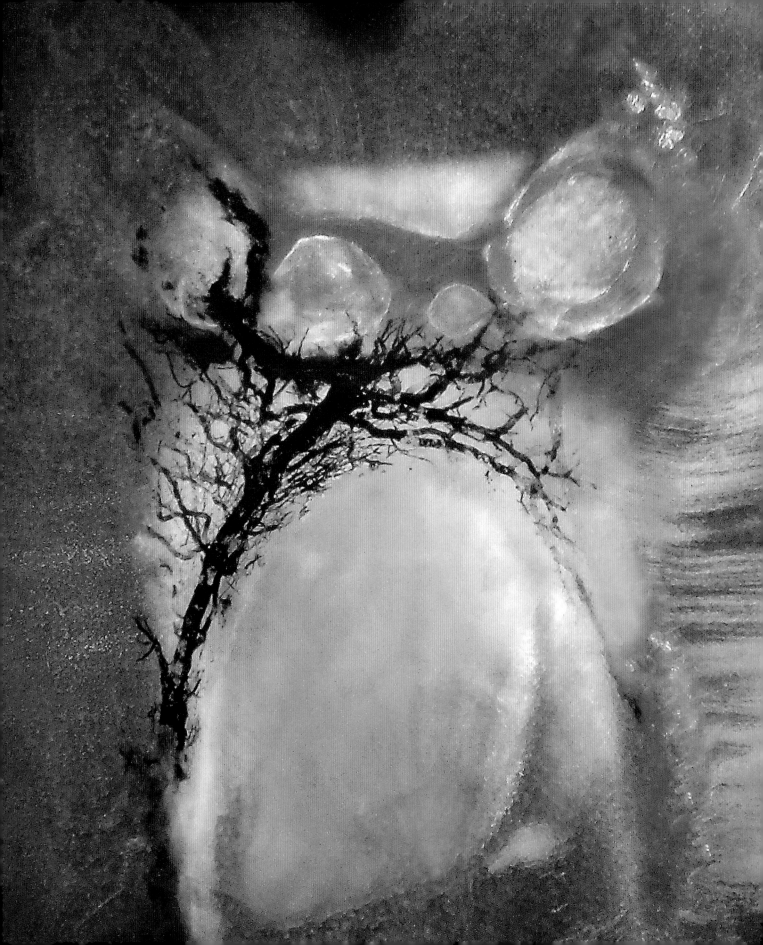

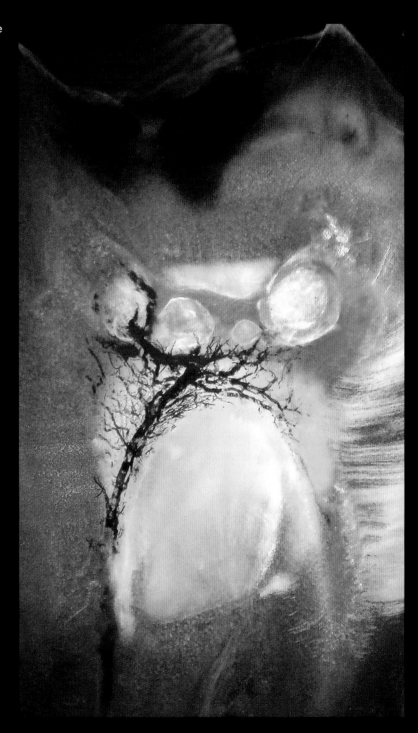

FIG 7-8 Four pulp stones in the pulp chamber of a mandibular tooth. Irreversible pulpitis is demonstrated.

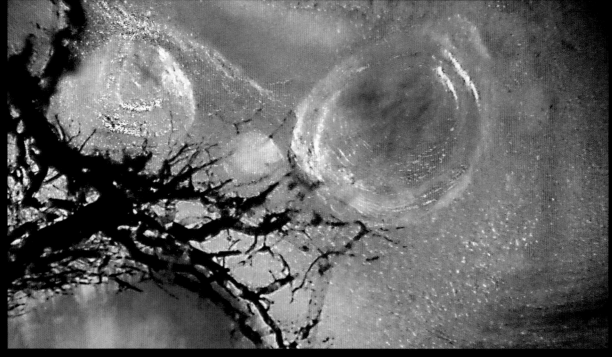

FIG 7-9 Closer images of the pulp horn region plus three of the pulp stones.

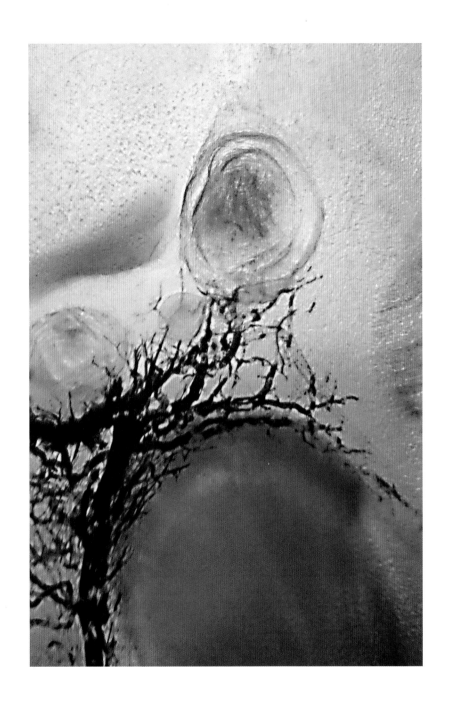

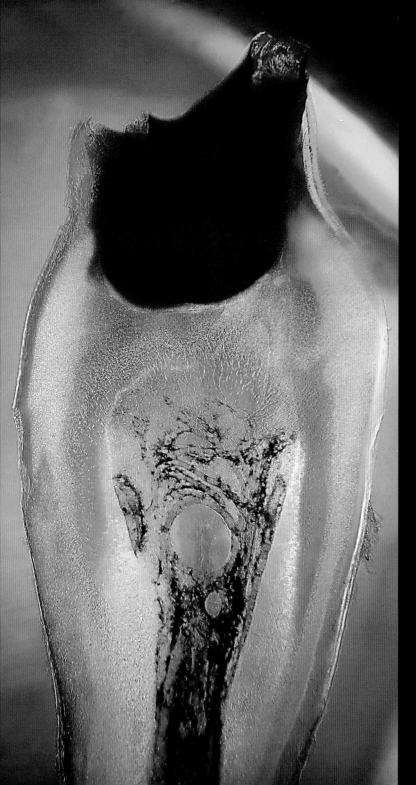

FIG 7-10 Another example of "vascular looping" in a mandibular premolar specimen. A composite restoration is present.

FIG 7-11 A result where the perimeter
vascularity is greater in volume than
the supply. It is my hypothesis that this
could be attributed to a drug side effect

8

Raschkow Plexus

The Raschkow plexus, also called the *nerve plexus of Raschkow*, is a network of myelinated nerve fibers. It is located in the pulp of the tooth between the core and the cell-rich zone and consists of bundles of axons that are unmyelinated but surrounded by Schwann cells and basal lamina. The terminal branches pass through the odontoblasts to synapse with the odontoblastic process within the dentinal tubule.

The Raschkow plexus monitors and transmits the sensation of pain from the pulp. This is done with several types of nerve fibers: Aβ- and Aδ-fibers and C-fibers. Myelinated A-fibers reach the odontoblasts, the predentin, and the dentin and extend into the dentinal tubules, establishing contact with the odontoblasts and conducting rapid and sharp pain sensations. Unmyelinated C-fibers, on the other hand, are thinner and terminate in the pulp tissue proper, and these are involved in dull aching pain. The odontoblastic layer is innervated by sensory nerve fibers originating from the inferior and superior alveolar nerves, which branch to form the subodontoblastic Raschkow plexus.

In late 2019, through dedicated effort, I was finally able to visualize the nerve. The nerve of a tooth has been defined and described in histologic sections but not in 3D diaphanization. This chapter still needs a lot of work, but it presents the plexus of Raschkow via diaphanization and in conjunction with the exposed vascularity.

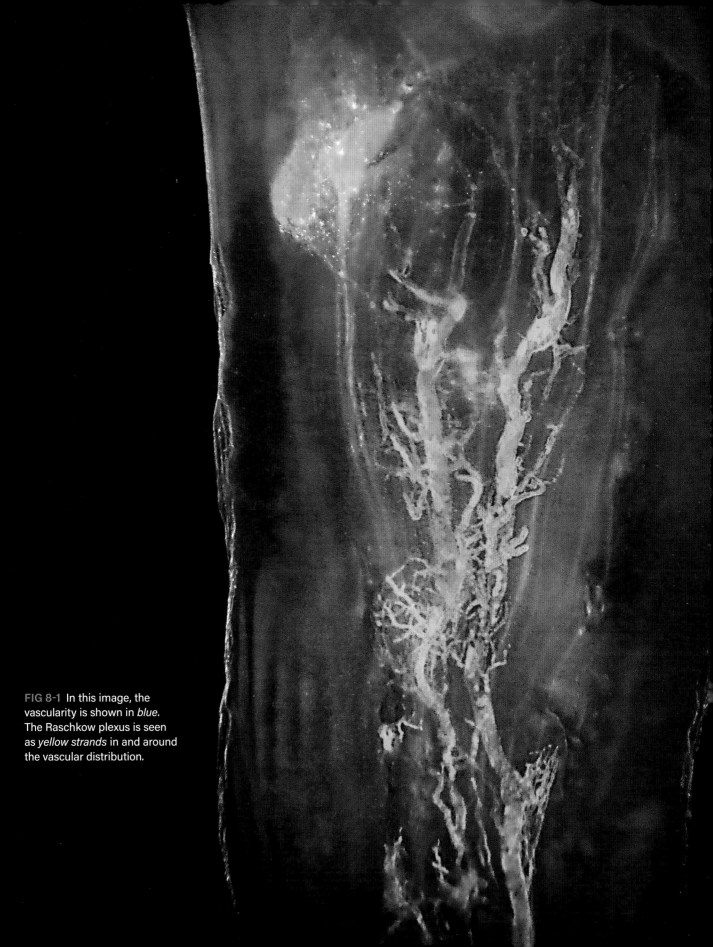

FIG 8-1 In this image, the vascularity is shown in *blue*. The Raschkow plexus is seen as *yellow strands* in and around the vascular distribution.

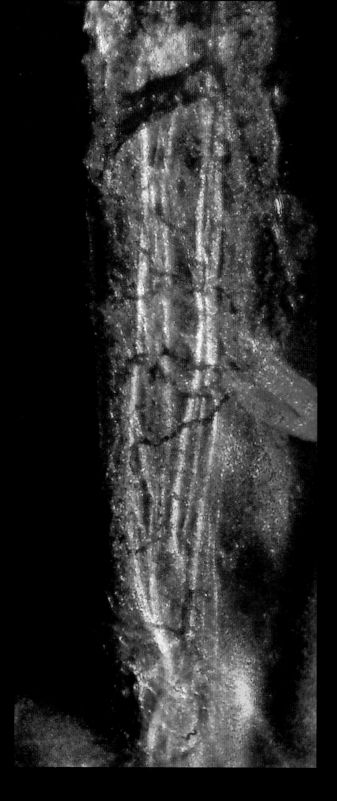

FIG 8-2 The *electric blue* is the Raschkow plexus. The vascular distribution can be visualized. On the recto, two more images of the sample are pictured. The vascular distribution is seen in the left-side image. The vascularity plus the plexus of Raschkow are both seen.

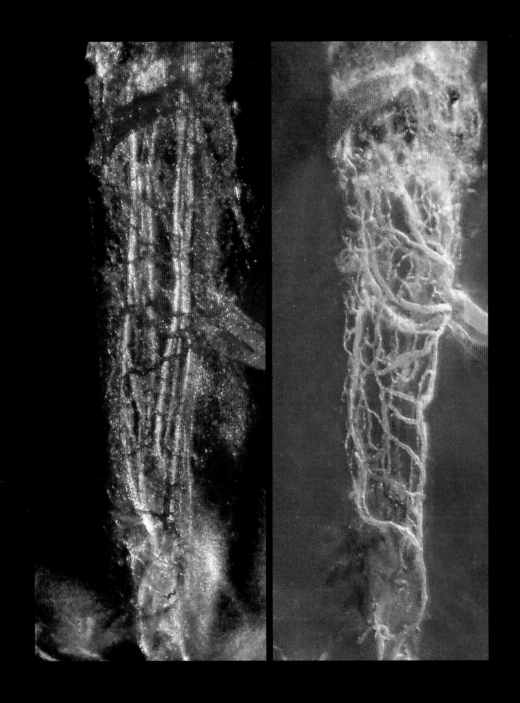

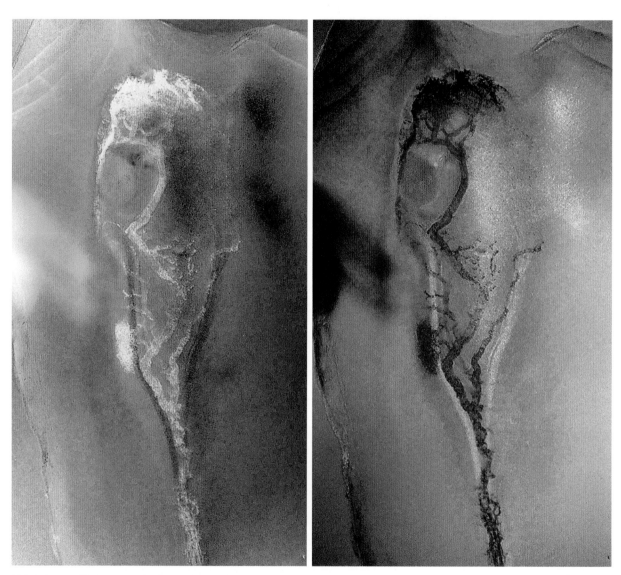

FIG 8-3 Mandibular premolar imaged in varied lighting. This tooth demonstrates late-stage irreversible pulpitis to early necrosis. Branches of the plexus of Raschkow can be visualized.

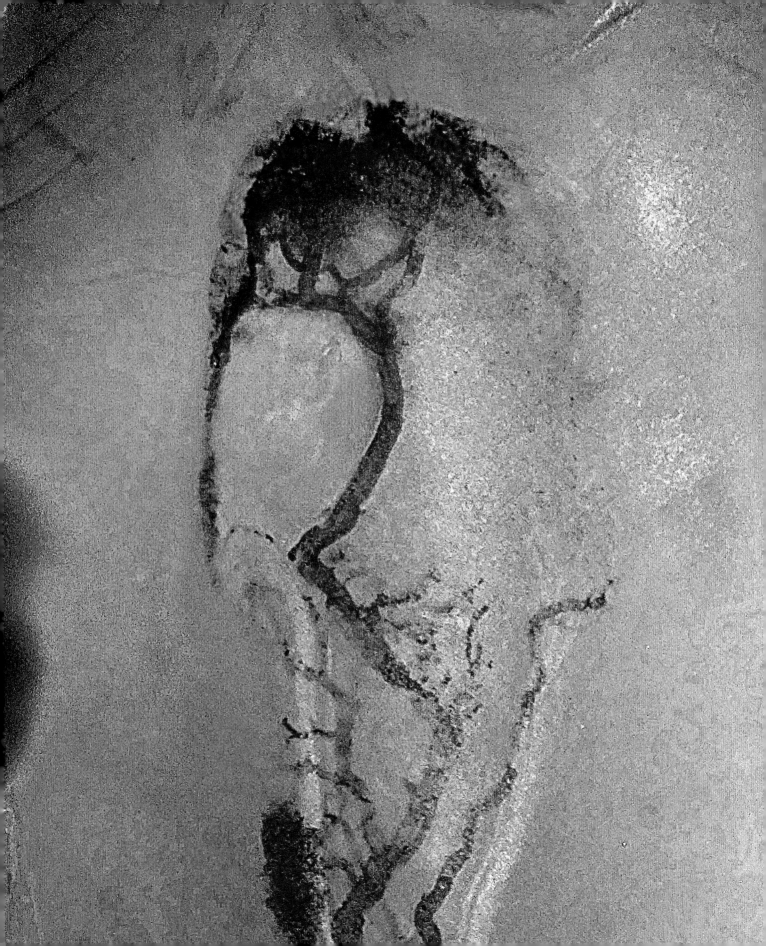

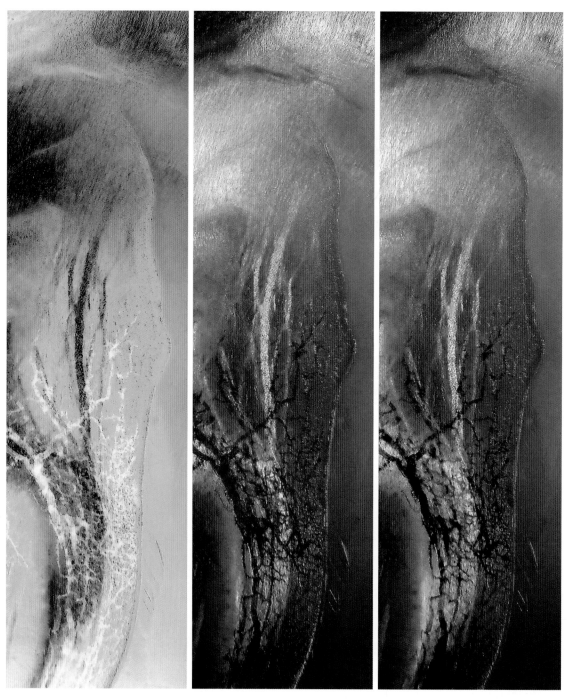

FIG 8-4 Images with varied lighting of a maxillary molar. The vascularity plus the plexus of Raschkow can be seen in the pulp horn of the pulp chamber.